RADIATION PROTECTION STANDARDS

Author
Lauriston S. Taylor
President, National Council on
Radiation Protection and Measurements
Washington, D.C.

LONDON

Butterworths

THE BUTTERWORTH GROUP

ENGLAND

Butterworth & Co (Publishers) Ltd
London: 88 Kingsway, WC2B 6AB

AUSTRALIA

Butterworth & Co (Australia) Ltd
Sydney: 586 Pacific Highway, 2067
Melbourne: 343 Little Collins Street, 3000
Brisbane: 240 Queen Street, 4000

CANADA

Butterworth & Co (Canada) Ltd
Toronto: 14 Curity Avenue, 374

NEW ZEALAND

Butterworth & Co (New Zealand) Ltd
Wellington: 26-28 Waring Taylor Street, 1
Auckland: 35 High Street, 1

SOUTH AFRICA

Butterworth & Co (South Africa) (Pty) Ltd
Durban: 152-154 Gale Street

This book represents information obtained from authentic and highly regarded sources. Reprinted material is quoted with permission, and sources are indicated. A wide variety of references is listed. Every reasonable effort has been made to give reliable data and information, but the author and the publisher cannot assume responsibility for the validity of all materials or for the consequences of their use.

Copyright © The Chemical Rubber Co. 1971
First Published 1971
Printed in the U.S.A.
ISBN 0 408 70327 X

CRC MONOSCIENCE SERIES

The primary objective of the CRC Monoscience Series is to provide reference works, each of which represents an authoritative and comprehensive summary of the "state-of-the-art" of a single well-defined scientific subject.

Among the criteria utilized for the selection of the subject are: (1) timeliness; (2) significant recent work within the area of the subject; and (3) recognized need of the scientific community for a critical synthesis and summary of the "state-of-the-art."

The value and authenticity of the contents are assured by utilizing the following carefully structured procedure to produce the final manuscript:

1. The topic is selected and defined by an editor and advisory board, each of whom is a recognized expert in the discipline.

2. The author, appointed by the editor, is an outstanding authority on the particular topic which is the subject of the publication.

3. The author, utilizing his expertise within the specialized field, selects for critical review the most significant papers of recent publication and provides a synthesis and summary of the "state-of-the-art."

4. The author's manuscript is critically reviewed by a referee who is acknowledged to be equal in expertise in the specialty which is the subject of the work.

5. The editor is charged with the responsibility for final review and approval of the manuscript.

In establishing this new CRC Monoscience Series, CRC has the additional objective of attacking the high cost of publishing in general, and scientific publishing in particular. By confining the contents of each book to an *in-depth treatment* of a relatively narrow and well-defined subject, the physical size of the book, itself, permits a pricing policy substantially below current levels for scientific publishing.

Although well-known as a publisher, CRC now prefers to identify its function in this area as the management and distribution of scientific information, utilizing a variety of formats and media ranging from the conventional printed page to computerized data bases. Within the scope of this framework, the CRC Monoscience Series represents a significant element in the total CRC scientific information service.

B. J. Starkoff, President
THE CHEMICAL RUBBER CO.

This book originally appeared as part of an article in *CRC Critical Reviews in Environmental Control*, a quarterly journal published by The Chemical Rubber Co. We would like to acknowledge the editorial assistance received by the journals' editors, Prof. Richard G. Bond and Dr. Conrad P. Straub, both at the University of Minnesota. Mr. Harold O. Wyckoff, Armed Forces Radiobiological Research Institute served as referee for this article.

AUTHOR'S INTRODUCTION

It is now some 75 years since x-rays and the radiations from radium were discovered and almost overnight put to use in the service of man. It was only a matter of weeks after their discovery that their potential for injury was also observed and rudimentary precautions against their ill effects instituted. But at the same time, the almost fantastic benefits from the uses of ionizing radiation were so obvious that the rapidity in the growth of applications outstripped the basic knowledge of the physical and biomedical characteristics of the radiation itself. As a result, a number of radiation-caused injuries became evident and by the second decade, organized steps began to be undertaken to reduce undesirable radiation effects.

The report to follow is designed to be an outline of a case history of how an agent, beneficial to mankind, yet dangerous also, may be harnessed and controlled to a degree commensurate with societal needs and with the other risks to which all man is exposed. The problem of striking a compromise between need and risk is by no means unique to radiation; it is just that radiation has been studied longer and more exhaustively than almost any other hazardous — yet essential — agent that comes to mind. Part of the difficulty is the apprehension about radiation that is often out of proportion to all the other risks we are willing to live with. The approach used in solving the problems of protection against ionizing radiation could indeed provide a model for the control of other harmful agents which, while with us for many years, are just now being confronted.

In 1915 the first proposals for radiation protection recommendations were made by the British. Still lacking a unified means of measurement, it was not until the early 1930's that quantitative control of radiation was possible. In the meantime, the rudiments of an understanding of the biological effects of radiation were being acquired and, by 1940, early systematized schemes for radiation protection were in use for x-rays and gamma rays. Following the advent of atomic energy, when other radiations and radioactive materials became available in great quantity, the need for protection measures became much more widespread. Various national and international organizations became engaged in the development of radiation protection recommendations, codes, and regulations to the point where radiation today is probably the most-studied, the best-understood, and the best-controlled environmental pollutant.

This report traces the development of the understanding of radiation hazard, and the philosophy for protection against it.

It is a case history, also, of the effectiveness of organized group actions on a national and international scale. While drawing on the works of many individuals and organizations, special attention is given to the work of the National Council on Radiation Protection and Measurements (NCRP) and the International Commission on Radiological Protection (ICRP) by which most of the basic radiation protection philosophies and criteria in the world have been generated.

The author is indebted to the following associates who have checked appropriate parts or all of the manuscript: F. D. Sowby (ICRP), Paul Tompkins, Federal Radiation Council (FRC), Francesco Sella, United Nations Scientific Committee on the Effects of Atomic Radiation (UNSCEAR), Harold O. Wyckoff (NCRP), International Commission on Radiation Units and Measurements (ICRU), Bo Lindell (ICRP), C. L. Dunham, National Academy of Science (NAS), Russell Morgan, National Advisory Committee on Radiation Protection (NACOR), John Villforth, Bureau of Radiological Health (BRH) and Jack Greene, Office of Civil Defense (OCD).

THE AUTHOR

Lauriston S. Taylor's principal education was at Stevens Institute of Technology and Cornell University where he majored in physics and where for three and a half years he was a Heckscher Research Fellow. His early researches were in the field of x-ray crystal structure, electronics and x-ray absorption spectra. He holds honorary degrees of Doctor of Science from the University of Pennsylvania and St. Procopius College.

Because of his x-ray work at the Bell Telephone Laboratories in 1922 and subsequent researches at Cornell, he was asked, in 1927, to organize an X-ray Standards Program at the National Bureau of Standards, at which time he accepted a position as the Head of an X-ray Group in the Atomic Physics Section. This program developed steadily and in 1941 he was made Chief of a new X-ray Section.

It was during this period that the National Bureau of Standards, under his direction, developed the first National Standards for X-ray Dosimetry in the United States. In 1931, he carried out the first direct comparisons of x-ray standards between the United States, England, Germany and France.

In 1948 he was on loan to the Atomic Energy Commission for the purpose of organizing and heading a Biophysics Branch in the Division of Biology and Medicine. It was during this period that he organized "Project Gabriel" to evaluate the long-range implications of strontium 90 in fallout.

In 1962, Dr. Taylor was made an Associate Director of the National Bureau of Standards. In 1965 he resigned to take a position with the National Academy of Sciences as Special Assistant to the President and Executive Director of their Advisory Committee to the Office of Emergency Preparedness.

Over the years, Dr. Taylor has published over some 130 papers and parts of 16 books related mainly to various phases of x-ray protection and x-ray measurements. He has also received many national and international awards and citations for his work in these fields.

In addition to his formal occupation at the present time, he is the President of the National Council on Radiation Protection and Measurements, Chairman Emeritus of the International Commission on Radiation Units and Measurements, Member Emeritus of the International Commission on Radiological Protection, and a member of numerous other advisory or special technical committees.

TABLE OF CONTENTS

The Tolerance Dose	13
Advisory Committee on X-ray and Radium Protection (1928)	15
Advisory Committee on X-ray and Radium Protection	17
Radium Protection for Amounts up to 300 Milligrams (Report No. 2)	18
X-ray Protection (Report No. 3)	19
Radium Protection (Report No. 4)	19
Radioactive Luminous Compounds (Report No. 5)	19
International Activities	20
Manhattan District	21
Other Protection Activities (1940's)	22
The National Committee on Radiation Protection	23
NCRP Report on External Expsure	24
NCRP Report on Internal Emitters	28
NCRP Report on X-ray Protection	30
NCRP Reports on Handling and Disposal of Radioactive Materials	31
NCRP Reports No. 8, 9, 12 and 16	32
NCRP ReportsNo. 10	35
Exposure of the Public	35
Atomic Bomb Casualty Commission	36
Radiation Quantities and Units	37
Reconstitution of the International Committees	37
NCRP Activities 1953 to 1956	41
NCRP Report No. 19	43
Other Developments 1950 to 1955	45
1956 – A Year of Changes	47
ICRP Meetings – (April, 1956)	48
Change of MPD by NCRP – 1956	49
ICRU – ICRP Study for UNSCEAR	50
Public Health Service – 1952	52
Congressional Concern Over Radiation Hazards	52
Federal Radiation Council (FRC)	53
NCRP Relations With NBS	55
Public Health Service Activities	56
National Academy of Sciences	56
Federal Radiation Council	57
Ad Hoc Report on Somatic Effects of Radiation	58
NCRP Reports Up to 1961	60
NCRP Report No. 20	61
NCRP Reports No. 21-28	61
Radiation from Television Receivers	64
JCAE Hearings	64
Changes in Understanding of Genetic Effects of Radiation	72
NCRP Activites 1958 to 1964	73
NCRP Report No. 29	75
National Advisory Committee on Radiation (NACOR)	77
ICRP Acitivities 1958-1970	79
ICRP Report on Permissible Dose for Internal Radiation (1959)	80
ICRP Report on "Protection Against X-rays up to Energies of 3 MeV and Beta and Gamma Rays From Sealed Sources"	81

ICRP Report on "Protection Against Electromagnetic Radiation Above 3 MeV and Electrons, Neutrons, and Protons" 82
ICRP Report on "Handling and Disposal of Radioactive Materials in Hospitals and Medical Research Establishments" 83
"Recommendations of the ICRP as Amended in 1959 and Revised in 1962" 83
ICRP Report on Principles of Environmental Monitoring 84
Reports Made to the ICRP 85
ICRP Publication #8 "The Evaluation of Risks from Radiation" 85
Recommendations on Radiological Protection (1965) ICRP Publication #9 86
ICRP Publication #10 Evaluation of Radiation Doses to Body Tissues from Internal Contamination to Occupational Exposure 87
ICRP Publication #11 Radiosensitivity of Tissues in the Bone 87
ICRP Publication #12 On General Principles of Monitoring for Radiation Protection of Workers 88
ICRP Publication #13 Radiation Protection in Schools for Pupils up to the Age of 18 Years 88
ICRP Publication #14 "Radiosensitivity and Spatial Distribution of Dose" 89
NCRP Reports 1966 to 1971 90
Report on Neutron Protection #38 94
NCRP Review of Basic Radiation Protection Criteria (1959-1970) 94
Political Action Activities 97
United Nations Scientific Committee on the Effects of Atomic Radiation (UNSCEAR) 98

RADIATION PROTECTION STANDARDS

Author: **Lauriston S. Taylor**
Special Assistant to the President
National Academy of Sciences
Washington, D. C.

Referee: Harold O. Wyckoff
Defense Atomic Support Agency
Armed Forces Radiobiology Research Institute
Bethesda, Maryland

Ionizing radiation, as first recognized, was in the form of x-rays and radiations from radioactive materials. For the first two decades after its discovery, radiation uses and applications were mainly in the medical field in the hands of the physicians or of engineers who were working with physicians. It is not surprising, therefore, when some unusual physiological effects occurred to individuals working with these radiations that they were relatively quickly traced to the radiation itself as the causative agent. This is not to say that these cause-effect relationships were always accepted, and some of the earliest literature contains some interesting, and now amusing, theories as to the cause of some of the early observed effects.[1]

The published records concerning early efforts in radiation protection are scant and scattered and usually are buried as a part of some other subject. The earliest x-ray injury on record appears to have been observed in January 1896 by Grubbé and again in March 1896 when Edison reported that both he and his assistant were troubled by severe smarting of the eyes after several exposures to a "discharge tube."[2,3] By the end of the year 1896, skin dermatitis was definitely reported by several observers and the study of biological action of x-rays was soon the subject of numerous reports in the early literature.

However, radiation protection in its early development did not seem to follow any very orderly or systematic development along with other phases of the science of roentgenology. It seemed to develop in lulls and spurts - the latter brought on each time there was some relatively severe effect resulting from an obvious overexposure. It now seems probable that in the excitement over new discoveries being made almost daily with x-rays, protection from them was allowed to lapse after the first few meager precautions were taken. It should, in justice, be pointed out, though, that in those first few years there was little quantitative information on the absorption and scattering of x-radiation, and hence the efforts made toward protection were largely of a "cut and try nature" without any definite goal of achievement or, for that matter, any definite way for quantitatively defining the radiation.

The problem was somewhat complicated by the lack of unanimity of opinion as to the real caustive agent in the production of injuries. For example, one school of thought led by Tesla[4] believed the injury to be due to some kind of high frequency electrical phenomenon. Kassabian, in 1907,[1, 5] thought the causes of radiation burns might be: a. ultraviolet radiation, b. flight of minute target particles to the skin, c. scattered cathode rays, d. x-rays, e. electrical induction, f. ozone generation in the skin, g. idiosyncrasy, or h. faulty technique.

Of these, the burn by high-frequency electrical induction seemed one of the most difficult to disprove. (This latter is of interest because of our present knowledge today about the injurious effects of microwave radiation. While that was clearly not the cause of the early injuries the consciousness of "high frequency" radiation as an injurious agent is of note.)

As early as 1898, it was found that guinea-pigs placed in a double Faraday cylinder (thus shielding off any high-frequency radiation) were found to be injured by the radiation passing through the aluminum walls of the containers.[6] In 1901, it was found that the injury caused by the x-rays was not limited to the skin as in all earlier observations, but also appeared in depths within the body of the animals.[7] It was also observed that experimental doses sufficient to cause abortion were possible.

Within the first few months after their discovery it was found that x-rays were stopped more effectively by lead than any other common metal and, hence, lead came into very early use for radiation protective purposes. It was because of this observation, together with the absence of any knowledge as to radiation dosimetry, that all of the early radiation protection recommendations, to be referred to later, were in terms of shielding factors rather than dose factors.

As contrasted to conditions today, the need for protecting the patient during radiographic exposure was very great. This is easily understood when one considers that very low voltage was used on the tubes with the consequent long exposures to the very soft, and easily absorbed, radiation. For example, a radiograph of the spine might require exposures up to one and a half hours and skin burns were a frequent result. That the radiologist was not more frequently affected by scattered radiation from such exposures was attributable to the fact that during the exposures he might retire to another room to see other patients - or even go out for lunch.

While the presence of scattered radiation in the radiographic rooms was undoubtedly known from the first,[8] attention was first drawn to its possible danger in about 1903 when Rollins pointed out,

"As x-light spreads its spherical wave, filling a room with an actinic mist which radiated x-light in every direction....it was essential to prevent the escape of every ray not needed to make the photograph or to illuminate the screen."[9] At about this time there appeared a multitude of protective devices to be worn by the radiologist: the apron, jacket, gloves, goggles, and so on until his actions were so restricted that of necessity other means of protection were sought. Then at about the time of the First World War, the more elaborate array of individual protective devices began to disappear in favor of protection built into the apparatus. During all of this period, there was an almost complete lack of quantitative knowledge about exact radiation measurements.

The evolution of radiation protection was greatly assisted by the radiologists' endeavor to establish a maximum dose which could be tolerated, more or less continuously, by the human body. However, until the 1920's the dosimetry was almost entirely in terms of changes in physical and chemical effects on materials including ionization, photographic effects, photovoltaic effects, photochemical effects, etc. Such approaches were mostly doomed to failure because the indicators did not have adequate sensitivity and depended on the purity of materials. Furthermore, their response as a function of energy was not determined or understood. One of the first attempts to establish a radiation tolerance was in 1902 when it was proposed that if the photographic plate was not fogged in seven minutes of exposure, the radiation would not be of harmful intensity - presumably for nearly continuous exposure.[11] It was not until the late 1920's that there was any international agreement on a quantity and unit for the measurement of radiation dose which made it possible to uniformly describe and define radiation exposure.[10]

Following a paper on protective devices read at a meeting of the British Roentgen Society in June 1915,[12] what is believed to be the first organized step toward radiation protection was made. At that meeting the following resolution was introduced by a Mr. C. C. Leyster, and was unanimously adopted:

"That in view of the recent large increase in the number of roentgen-ray installations, this Society considers it a matter of the greatest importance that the personal safety of the operators conducting the roentgen-ray examinations should be secured by the universal adoption of stringent rules and that the Council of this Society be requested to meet at an early date to take steps to secure this."[13] It is significant that this was practically the last mention of radiation protection for a period of nearly five years, due partially to the press of a world war and partly to the human characteristic of indifference.

Attention was next focused on radiation hazards in about 1920, following a deluge of newspaper publicity over the deaths here - and abroad - from aplastic anemia, especially among individuals who had worked with the military medical services. It was also during this period that the hot-cathode x-ray tube came into common use and delivered quantities of radiation many times greater than possible with the earlier gas-type tubes. Nevertheless, the publicity stirred many countries into belated action and, once begun, workable safety precautions began to be established. Possibly the first standing radiation protection committee was formed by the American Roentgen Ray Society in September 1920.[14] At about the same time a cooperative committee was formed between several medical and radiological groups in England and this committee published what is probably the first general set of radiation protection recommendations in the Journal of the Roentgen Society in July 1921.[15,16] The recommendations of the American Roentgen Ray Society were later adopted in their annual meeting in 1922. In substance the British and the American recommendations were very much the same, although the former were in more detail.

In tribute to those two original committees, it should be pointed out that the fundamentals of radiation protection as put forth in their reports remained essentially unaltered in the decade that followed. By the end of 1922, it was found that standards for radiation protection had been set up and accepted by many countries. By the middle 1920's, most of the more technically advanced countries had developed some type of protection philosophy and protection recommendations.

The way having been indicated, together with a simple pattern for radiation protection recommendations, many other countries began developing their own recommendations. Of these, perhaps the Swedish were the most active, but substantial work was also done by such countries as Holland, Germany, Soviet Union, and France. Since there were no important differences between these and the British proposals they will not be dealt with further here.

At the First International Congress of Radiology, held in London in 1925, the need for recommendations for radiation quantities and units was discussed at length, and as a result an ad hoc group was formed to study the units problem and to report at the next Congress being scheduled for Stockholm in 1928. The need for more comprehensive recommendations for radiation protection was also discussed, but apparently it was felt that there was not enough information available at the time upon which to formulate a plan for future action. It was agreed, however, that that question would be brought up at the Stockholm Congress.

In part because of this action, a considerable amount of effort was put into the development of radiation protection philosophy and recommendations in a number of countries, seemingly in preparation for presenting positions in Stockholm. Of the several countries having relatively comprehensive recommendations to present at that Congress, the two best sets appeared to have been developed by England and Sweden. As will be noted below, the British proposals were to form the pattern for international adoption.[17]

Perhaps the most significant step made in the international recommendations was the introduction of specific tables of the lead thicknesses necessary to shield against radiations produced at given voltages for x-rays or for given quantities of radium. In this connection, it is interesting to notice that because of the lack of a suitable means for radiation measurements, none of the specifications included the dose rate of the beams being sheilded or distances between radiation sources and the operator. The specification was based on the requirement that it should not be possible for a "well rested eye of normal acuity to detect, in the dark, appreciable fluorescence of a screen placed in the permanent position of the operator." When one considers the relatively high dose rate necessary to produce fluorescence with the screens available in that day, it is fairly clear that the exposure levels of personnel in the early days could have been very high. Unfortunately, there is no way of knowing whether the levels were approaching an upper margin of "safety." It is therefore interesting to note in retrospect that since the 1928 shielding requirements were introduced and implemented, the statistically detectable injury to radiologists has been extremely small - if it exists at all.

While some groups were mainly concerned with the problems of direct shielding of radiation as the basic means of protection, a few others were trying to obtain a better understanding of the radiation injury mechanism and to establish levels of radiation exposure that might be regarded as acceptable for general use.

It is important to appreciate the development of the protection concepts and philosophy in the mid 1920's because within the next decade they had substantially developed and matured and provided the basis for the pattern which exists today. They also introduced concepts and terminology, some of which have since been modified, but others of which have hung on to cause confusion which persists even today in the basic understanding of some aspects of our radiation protection criteria.

The essentials of the philosophy were explained well by Wintz and Rump in a report to the League of Nations in 1931 and the discussion below is based to a considerable extent upon their summary.[18] In the 1920's it was necessary in gauging the injurious effects of radiation on the human organism to refer, by way of comparison, to the actions of poisons and to base their approach on the one normally used in toxicology. In this, the most important thing to know is what quantities of toxic agent may be harmful and what quantities can certainly be tolerated by the body without ill effects. Presumably, standards derived in this way may also be applied in the case of radiation.

In the therapeutic application of radiation the concept of a tolerance dose was important. It signified the amount of radiation which the tissue concerned was able to tolerate. Experience had already shown, however, that such a dose cannot be regarded as entirely harmless; for even though after its administration the tissue undoubtedly recovers from some of the morphological alterations which were known as radiation injuries, the tissue is nevertheless "left with a reduced power of resistance to otherwise inoperable influences." Further investigations have shown that this *locus minoris resistentiae* is engendered even by doses equivalent only to 50% of the tolerance dose. This was felt to hold good for average tissue even though the radiosensitivity, and therefore the tolerance dose, might vary some for each kind of tissue.

These early observations showed that the tolerance dose was never a harmless one; therefore, tolerance doses could in no cases be readministered indefinitely to any particular part of the tissue after the visible effects have disappeared in each occasion. The only dose, therefore, that could be harmless would be one that would have no effect whatever on the particular tissue concerned and, hence, be administered *ad libitum.*

It is interesting to note here that these general concepts, based somewhat on experience and somewhat on assumptions, dealt with a cause-effect relationship which could not be positively proven or disproven. Precisely the same situation exists today, although currently our knowledge is much more extensive and our means of measurement and definition much more refined.

It was thus recognized that the conditions to be filled by such a "tolerance dose" were very stringent and would seem to warrant the question as to whether such a dose can exist at all.

It was understood that when radiation interacted with living cells part of the energy was absorbed, but the only elementary process which was known with certainty would occur when x-rays impinged upon a material object was the ejection of electrons of greater energy than those normally in the atom. When the atom reverted to a condition of its original energy, a form of radiation known as fluorescent radiation occurred. A collision of the electrons might also release energy in the form of kinetic energy or chemical activity. If a large part of the radiation was absorbed, there was an accession of energy among the numerous cell molecules and when this exceeded the amount compatible with life activity of the cell, the cell functions were disturbed or the cell died. However, since such an accession of energy might occur even on the absorption of a very small amount of radiation, there was also some belief that when only a few molecules were stimulated, the life activity of the cell might be intensified. This fell in line with earlier observations in the case of light and heat and explained some of the stimulating effects that were observed by numerous research workers, at that time, resulting from the administration of small amounts of x-rays (this question of a "stimulating effect" was not clear then and has not been fully settled even today).

Thus, the conclusion was reached that a really harmless dose of radiation could only be said to have been given if it was incapable either of destroying or damaging the cells beyond any stimulating action.

The exact determination of such a dose seemed to be impossible (it still is) and accordingly it was the custom to speak of a "harmless dose" whenever no alteration in normal condition and activity of the body could be detected by the available methods of medical examination and observation. This was the principle adopted in subsequent investigations undertaken with the object of determining the magnitude of a harmless x-ray dose.

Bearing in mind that during the 1920's moderately high voltage x-rays had come into relatively common clinical use (150 to 200 kV), not only was there soft radiation absorbed largely near the surface but also there was the hard radiation absorbed at depth and in considerable quantity. It had long been known that most organisms possessed the power of adaptation or toleration to external insult and it was reasoned that the body cells were similarly able to adapt to, or to tolerate, or to neutralize, small amounts of x-rays. This capacity of the cells for habituation under the administration of therapeutic amounts of radiation was clearly demonstrated by the effects of fractional doses. For example, it was observed that if 100% of a "skin unit dose" required to cause an erythema was distributed over a period of five days, instead of being given in approximately one half hour, the biological effect was about 35% less.

It was also observed that for therapeutic irradiation conditions with the extended distance between the patient and the source, the dose conversion could not be made by the simple inverse square law alone. The computed dose, to be equivalent to that at a shorter distance, had to be increased by an amount of radiation which was then termed the "additional biological dose." All of this seemed to "indicate that the body cells were able to deal rapidly with the energy of small amounts of x-rays and that a particular effect could not be secured by adding together the several doses given separately." The administration of a specific amount of radiation at one sitting thus produced a different effect from the gradual absorption of the same amount of radiation that might occur, say, in the case of radiation workers.

It was concluded that the comparison of biological actions was not strictly legitimate when the absorption of radiation is distributed over long periods of adaptation such that much of the radiation absorbed by the body may be wholly without biological effect.

It was considerations such as outlined very briefly above that led to the many early endeavors to establish what subsequently became known as a "tolerance dose" and later as a "permissible dose" and in other fields as "allowable" or "acceptable" doses. In all of these the basic concepts have been the same. Additionally, insofar as radiation is concerned, it should be clear from the philosophy just outlined that the adoption of tolerance or permissible doses did not depend upon the assumed existence of a threshold; on the other hand, the setting of the tolerance level was on the basis of whether or not effects could be observed, and this has been the common approach to similar problems for all other toxicological agents.

Therefore, it is rather interesting to realize that because radiation as a toxic agent has perhaps been studied longer than most other such agents, attempts at sophistication of the understanding of the problem have led to the adoption of concepts for which there are even today no complete biological bases.

THE TOLERANCE DOSE

Beginning in the early or mid 1920's, considerable effort was directed to attempts to establish some level of dose which it was felt might be acceptable to radiation workers. It was recognized that any studies would be conditioned by the technical limitations of measurement as well as by medical observations on the patient. It was also recognized that any numerical data would be of only conditional validity. Since reliable measurements became possible only in the late 1920's, efforts were made to go back to compute the amounts of radiation that had been received in earlier years by persons who had sustained observable injuries. It was felt that such reconstruction was almost impossible in most situations and that even the injuries themselves could not be clearly defined as having been caused by radiation. It was noted that the major injuries, which were those occurring to the hands, were probably not caused by radiation alone but by other agents as well, such as film processing chemicals or disinfectants. It was also recognized that the ozone and nitrous oxide present in most x-ray areas probably also had to be taken into account and that shielding of x-ray tubes and rooms was quite irregular so that while one side of a room might be thoroughly "safe" the other side of the room might be relatively "dangerous."

Hence, it was not until it became possible for the first time to almost or completely encase x-ray tubes that interpretable measurements were feasible. Consequently, it was extremely difficult to assign any specific numerical values to a tolerance dose based on observations of effect or no effect of radiation workers prior to about 1925 or 1930.

The first attempt to define a tolerance dose in terms of a numerical value was made by Mutscheller.[19] He based his calculations on several good installations where it would be possible to establish good work patterns, and hence exposure patterns, for the radiation workers. He found that in the course of a month the members of the staff had received what was described as about 1/100 of a threshold erythema dose (known as the "skin unit dose"). Based on a monthly work time of 200 hours it was concluded that radiation should be stated as harmless at that level: assuming there was no recovery it would require 20,000 hours or more of exposure to produce an erythema. On the basis of an eight-hour day it would require about eight years to reach this level.

It was recognized that the definition of tolerance dose was of limited practical value, even from the technical point of view of its measurement and evaluation. Additionally, it was recognized that because of the time factors involved and the "biological additional dose" required, the actual exposure over long periods of time could almost certainly be very much greater. It was estimated that allowing for prolonged and continuous irradiation, a time factor value of at least 10 could be applied essentially as an additional safety factor. It was noted that even under conditions of more adequate protection than had usually prevailed in the past, the organism received a dose a hundred times as great as the tolerance level yet the skin was never visibly affected. Some early attempts to check this were made by irradiating mice at levels of 6 to 60 roentgens for an hour each day on the same skin field, and yet no discernible skin reactions were observed even after a considerable time.

In endeavors to use Mutscheller's concept, it was considered best to express the skin unit dose (the amount of radiation which, with the short-time method of administration, produces a definitely harmless skin reaction) in roentgens. A poll among a number of radiotherapists yielded an average value for x-radiation as currently used in deep therapy wherein one skin unit dose equaled 550 R, not counting the additional dose due to backscattered radiation.[20] This was rounded out to roughly 600 R and there began to be some general acceptance of Mutscheller's method of defining a tolerance dose. At the same time, because the time factor was so uncertain, it was agreed to disregard it because any uncertainty introduced thereby would be in the safer direction. Under these conditions the tolerance dose worked out as 1/100 of a skin unit dose or 6 R per month, or assuming 200 working hours per month about 10^{-5} R/sec.

The actual numbers arrived at during this period are of interest mainly for their general, rather than absolute, nature. As a matter of fact, measurements were made in terms of roentgens or other units of various size and the relationships between them was in considerable uncertainty. For example, there was also the Solomon R-unit, which had no really definable relationship to the roentgen - except that it involved ionization measurement.

Various other workers made estimates along the same general lines, and while most of these were based on a very small number of cases each, they arrived at a surprising degree of agreement. For example, Glocker and Reuss, and Behnken arrived at the same value.[21,22] Bouwers and Van derTuuk arrived at a figure of 0.7×10^{-5} R/sec.[23] Barclay and Cox reported on a case in which 0.007 skin unit dose per day was tolerated by an individual for six years without ill effects.[24] On the basis of an 8-hour day this would have represented a dose of 15×10^{-5} R/sec. Chantraine and Profitlich computed the dose at 0.5 to 0.1 R/hr which would reduce to 1.4×10^{-5} R/sec.[25] Solomon assumed that 4000 R (French roentgens) per year was harmless;[26] on the basis of 300 eight-hour days this came to about 7×10^{-5} R/sec. Schechtmann felt that the tolerance dose did not exceed about 1 R per month because he believed that the ionization in the air in the radiation area would not be without some contributary importance.[27] Fricke and Beasley and Jacobson made direct measurements in various clinics in Cleveland and New York and found in those areas that no exposure exceeded about 25 R/yr which again reduced to about 0.3×10^{-5} R/sec.[28,29] In examining the people so exposed they were unable to detect any blood changes even in the case of one radiologist who had worked for years without a lead rubber apron and who, at one time, was exposed to about 5 R/hr. The experience of Wintz led him to believe that 10×10^{-5} R/sec could be regarded as "safe". However, since he felt that for such purposes one should not operate at what might be a marginal level, he recommended 1×10^{-5} R/sec as a tolerance dose based on 300 eight-hour working days per year.[18] At the same time he suggested that in the case of persons who remained close to sources of continuous irradiation as, for example, radioactive preparations, the tolerance dose should be proportionately reduced and suggested that about $1/3 \times 10^{-5}$ R/sec should be regarded as harmless.

It is important to have gone through the exercise above for two principal reasons:

1. A crude evaluation and summation of all of the above results was what led to the adoption of the tolerance dose of 0.1 R per day by the NCRP and 0.2 R per day by the ICRP in 1934, and

2. All of the above evaluations were based on an absence of any kind of observable effect and hence were very probably on the conservative side as far as protection purposes are concerned, because the statistics on the erythema dose were very poor. (There were cases where there was damage but it was thought that these could be traced to excessive exposure at some time in the past.)

INTERNATIONAL COMMITTEE ON X-RAY AND RADIUM PROTECTION (1928)

The ICRP was formed in 1928 under the auspices of the Second International Congress of Radiology then meeting in Stockholm. The formation of an international protection committee had been discussed by the First Congress in 1925 but no active steps had been taken at that time toward this end.[30,31]

During the Second Congress, initial plans for organizing an international protection committee were developed by G. W. C. Kaye and S. Melville (Great Britain), and L.S. Taylor (U.S.). Dr. G. Grossman (Germany) and Dr. R. Sievert (Sweden) were brought into the discussions and plans were completed for the organization of a committee. Their proposals and organization plans were presented to the Second International Congress and were approved. The committee was named the International X-ray and Radium Protection Committee, and its membership consisted of the five individuals who planned the organization.

Because of the early experience of the Units Committee with its large and unwieldy membership, the Protection Committee decided at the outset to keep its membership as small as possible. It also decided to include in its membership only people who were working actively in the general field of radiation protection. For this reason, the initial Committee consisted of only five persons. This was enlarged for the next Congress by adding Dr. I. Solomon of France and Dr. E. Pugno-Vanoni of Italy.

The ICRP, because of its small size, was operated much more informally than the ICRU. The Honorary Secretaries from 1928 through 1937 were Dr. G. W. C. Kaye, then of the National Physical Laboratory, and Dr. Stanley Melville from 1928-31. It might be remarked (with some interest in view of later events) that at the time of its formation the problem of radiation protection was not considered to be as important or as serious as that of radiation units. Members of the ICRP felt that the problem would grow and sought to be prepared.

The Committee held its initial meeting in 1928 for the purpose of adopting some interim protection regulations. Until that time, the only clearly formulated recommendations were those prepared a few years earlier by the British Committee.[17] These were used as a basis for discussion. The first recommendations of the Committee were very similar to the early British proposals,[32] and until 1950 its recommendations were patterned around these early British recommendations.

The second meeting of the Committee took place in Paris in 1931 and was attended by all seven members.[100] At this meeting, there was some discussion of the possibility of introducing so-called tolerance levels of radiation for radiation workers. This was recognized as desirable, but with very little evidence to go on, no specific recommendations were made.

The 1928 recommendations were divided into five sections:

1. **Working hours**—This section included some working hour limitations not generally in vogue as of that period; for example, not more than five 7-hour days a week and not less than 1-month holidays per year. At the time the thinking leading to such recommendations was that daily working time limitations would serve to stretch out any exposure and consequently any effects that might develop. The extra vacation was designed to permit individuals to "recover" from any adverse effects of the radiation although none had ever really been observed. It had been noted that some radiation workers showed signs of fatigue and occasionally other systemic difficulty and it was not unreasonable to suggest extra time for a possible recovery. (However, it later began to appear that the early difficulties were probably more likely due to the ozone and nitrous oxide and generally poor working conditions which prevailed. By the end of the next decade the concept of special vacation and working hours for radiation workers had largely disappeared and it is now generally recognized that radiation workers have no need to be treated differently from other workers as long as the other physical requirements for protection are properly carried out.)

2. **General recommendations** (having to do with provision of light, ventilation and working temperatures)— This was again due to the need for removal of nitrous oxides and ozone. Additionally, it was frequently the pattern to put an x-ray department into any unused space that could be found and this was very apt to be some corner of a basement or other spot not generally considered suitable for medical work.

3. **X-Ray Shielding**— The x-ray shielding recommendations consisted of a table of lead thicknesses necessary to shield from radiations generated at different voltages from 75 to 225 kv. It also included precautions for protection against scattered radiations. (Prime attention was given to the protection of the radiation workers and in the early reports no mention was made of the patient.)

4. **Electrical Precautions**— During this period, air insulation was used for the high voltage and as fairly high voltages were coming more into vogue, there were probably more injuries from high-voltage than there were from radiation.

5. **Radium Protection**— This section included radium protective recommendations and here emphasis was put on working with shields, maintaining distance between operators and source, and moving with speed and care. Special attention was directed to the manipulation of radium emanation. Shielding was specified in terms of thickness of lead for a given quantity of radium element without regard to time and distance factors.

The third and fourth meetings of the International Committee took place in 1934 and 1937.[101,102] The meeting in July 1934 was notable because the Committee recommended a value of 0.2 R/day for the tolerance dose. (A value of 0.1 R/day had been recommended in the U.S. in March of that year.) This was based on the kind of evidence outlined above and supported by continued measurements in various institutions using instruments calibrated in the then accepted international roentgen.

No biomedical discussion or report was presented in explanation of the value adopted. Their statement was simply, "The evidence available at present appears to suggest that under satisfactory working conditions a person in normal health can tolerate exposure to x-rays to an extent of about 0.2 international roentgens (R) per day. On the basis of continuous irradiation during a working day of seven hours this figure corresponds to a dosage rate of 10^{-5} R/sec. The protective values given in these recommendations are generally in harmony with this figure under average conditions. No similar tolerance dose is at present available in the case of radium γ-rays."

The closing statement of uncertainty about a tolerance level for γ-rays indicated the continued uncertainty as to the basic biological mechanism involved. In the recommendations of 1937, the limitation relevant to radium γ-rays was omitted—again, without supporting discussion.

Throughout the period of 1928 through 1937, the main work of the International Committee consisted of gradually enlarging and extending the recommendations originally developed at its first meeting. The main item of the protection recommendations was the specification of the thickness of barriers of various materials that should be interposed between a source and an individual to assure adequate protection. Prior to the establishment of the permissible dose level in 1934 these barrier thicknesses had been somewhat arbitrary. In 1934 and 1937 data on the dose in Roentgens as a function of the barrier thicknesses and distance were put on a sound quantitative basis.

ADVISORY COMMITTEE ON X-RAY AND RADIUM PROTECTION

The roots of what is now known as the National Council on Radiation Protection and Measurements (NCRP) go back to 1928 and are intimately related to the formation of the International X-ray and Radium Protection Committee in July of that year. With the possibility in mind of forming an international organization on radiation protection, the Second International Congress of Radiology, before meeting in Stockholm in 1928, had invited several countries to send representatives to the Congress for the purpose of discussing protection problems and possibly preparing some initial x-ray protection recommendations. Because the United States radiological organizations had reached no agreement among themselves regarding any such recommendations, it was not possible to present a unified position from this country. There appeared to be somewhat similar problems among other groups as, for example, Germany, as to who carried the necessary authority to speak for the German radiologists.

As it turned out the only recommendations of significance that were offered for discussion were those of England and Sweden, and since the British were somewhat more detailed they were the ones, as noted above, that were adopted.

Because of the confusion, such as introduced by the United States and Germany, the honorary Secretary of the Committee, Dr. G. W. C. Kaye, made the informal recommendation that in those countries where there was more than one radiological organization, an effort should be made to establish a single central committee for the purpose of consolidating national recommendations for presentation at the next meeting of the Committee in 1931. It was suggested that the members take this up individually with the various groups in their respective countries. As the U. S. representative to the International Protection Group, it became the responsibility of L. S. Taylor to convey these recommendations to the various societies in this country and to convince them of its soundness.

In September 1928, the question was discussed informally with the officers of the American Roentgen Ray Society. Similar discussions were held with the president of the Radiological Society of North America and at a later time with the Radium Society. As a result of these discussions, the several organizations agreed to consolidate their protection activities into a single committee. To avoid any conflict between the societies, they also suggested that the Committee activities be centralized at the National Bureau of Standards for the following reasons:

1. By that time the NBS had established a definite long-range program in the general field of radiation protection.
2. The NBS was the only laboratory in the country having as its primary interest the development of radiation protection data and information.
3. The NBS had no inter-society or political ties and therefore could be expected to retain an independent position and viewpoint.
4. It provided an official U. S. representative to the International Committee on X-ray and Radium Protection.

The general organizational plan was to have each of the Societies recommend a physicist and a radiologist for membership. Because the development of protective measures was so intimately tied into the design and manufacture of x-ray equipment, it was also felt necessary to have adequate representation of the x-ray equipment and x-ray tube manufacturers.

Thus, early in 1929, an initial organization and program was established. The Committee was called "The Advisory Committee on X-ray and Radium Protection" and carefully avoided the question of to whom the Committee might be advisory. Actually, it felt itself to be responsive to the profession as a whole and clearly avoided any implied direct responsibility to either the government or to any individual radiological society.

This stratagem, while a source of minor misunderstandings at a later time, nevertheless was successful in that the committee quickly developed, and has always maintained, a reputation for objectivity and independence from any particular organization.

The original committee consisted of Dr. H. K. Pancoast and Dr. J. L. Weatherwax from the American Roentgen Ray Society; Dr. R. R. Newell and Dr. G. Failla from the Radiological Society of

North America; Dr. Francis Carter Wood from the American Medical Association; Dr. W. D. Coolidge and W. S. Werner from the X-ray Equipment Manufacturers; and L. S. Taylor, from the National Bureau of Standards and ICRP, acting as Chairman.

The first meeting of the advisory committee was held in September 1929 with an initial objective of preparing a set of detailed recommendations on x-ray protection.[55] This work was completed and published in May 1931 and was a tribute to the principle of what can be accomplished with a small committee.

The report covering some 25 pages was generally pragmatic in nature and did not attempt to go into the questions of the interactions of radiation and biological material. To a considerable extent the recommendations consisted of a very greatly expanded version of the 1928 recommendations of the International Committee. They did, however, take up some new areas of coverage and introduce some new principles.

For example, it is recognized that different kinds of apparatus, i.e., fluoroscopic radiographic, therapeutic, etc., had different operating requirements and as a consequence the equipment was divided into several classes, a pattern which has pretty much persisted up to the present. The presence of scattered radiation was also recognized and placed in perspective with different shielding rules for it as compared to the direct radiation.

The committee also recognized the special needs of the patient insofar as his exposure was concerned, and the first suggestions on patient protection appeared in this report. In all subsequent modifications of the report, protection of the patient was a point of emphasis with each new set of recommendations extending the preceding one.

Other sections included electrical protection, the use of x-ray equipment in anaesthetic rooms, and the storage of x-ray film, all of which were ancillary problems requiring attention. Another new area of concern, introduced for the first time, was the development of rules for technicians, nurses, and physicians in charge. Again, these have been enlarged upon as experience has been gained in subsequent years.

RADIUM PROTECTION FOR AMOUNTS UP TO 300 MILLIGRAMS (REPORT NO. 2)

The next effort was directed toward the preparation of a report on Radium Protection which was published in March 1934.[56] The committee established for this purpose included the original members, but added Drs. Sanford L. Withers and Curtis F. Burnam from the American Radium Society and Dr. L. F. Curtiss from the National Bureau of Standards.

This report was one of limited scope and dealt only with protection for radium in amounts up to 300 mg. However, this covered the great bulk of the radium usage as of that time and it was recognized that there were some greater uncertainties relating to larger amounts which would require longer-range study. It was essentially an enlargement of those portions of the international recommendation relative to radium protection and substantially more detail was added relative to manipulative procedures.

The report was notable for the amount of attention that it gave to working personnel for biomedical reasons, for example, the recommendation that permanent radiotechnicians have at least six weeks vacation a year, preferably four weeks during the summer and two weeks during the winter. In addition they were urged to spend as much time as possible out-of-doors during these vacations and after working hours. There were also strong recommendations for pre-employment physical examinations, blood counts, and the continued and frequent inspection of the hands of workers to detect any early evidence of exposure. It was indicated that blood counts at regular intervals of employment were highly desirable and, under proper conditions and wide interpretation, could be used as one of the safety criteria. Specifications were given for the level of white blood count below which the technician and working conditions should be subject to investigation.

Probably the most notable point in this report was the first quantitative recommendation of a tolerance dose for radiation workers. The report contained the statement, "The safe general radiation to the whole body is taken as 1/10 R/day for hard x-rays, and may be used as a guide in radium protection. Five R/day has been taken as the tolerance for the fingers. It must be emphasized that the calculation of radium dosages is not easy, and too great reliance is not to be put on the above figures."

This last statement was not to question the value of tolerance level of 1/10 R/day, but was to emphasize the fact that it was not then possible to control exposure by measurement means with very great accuracy. Because of this, some additional rules-of-thumb guidance were provided. These even covered such unusual conditions as holding a bare 100 mg radium preparation for five seconds—a procedure regarded as dangerous.

X-RAY PROTECTION (REPORT NO. 3)[57]

Soon after publication of the committee's first report in 1931, very rapid developments were made in the x-ray field; high voltage x-ray tubes operating up to 200 kVp, or more, became more reliable and into more common use in the ordinary hospital. In addition, experimental installations operating up to several hundred thousand volts were beginning to come into use. New developments began to make it possible to fully encase x-ray tubs in shielding containers and to more sharply delineate the beams to the field of interest and to shield the patient, radiologist, and technicians from undesired leakage radiation. Because of this, it was evident by 1934 that the earlier report would have to be revised and a new one issued in July 1936.

It might be worth noting that this handbook confirmed the specific permissible exposure level (then called tolerance dose) of radiation that could be allowed for occupational exposure. The figure recommended was the same as that proposed by the Advisory Committee in its radium report issued in March 1934 and was set at a level of 0.1 R/day. Parenthetically, it might be noted that this tolerance dose or permissible exposure level remained in force for some 12 years and was tested and used by the Manhattan District in its operations.[33,34]

RADIUM PROTECTION (REPORT NO. 4)

Following further study of the radium protection problems after the issuance of the report on protection from amounts up to 300 mg or more, an extensive set of recommendations was issued in August 1938.[58] This report dealt with the detailed problems of handling amounts of radium up to 1 g but, in addition, developed a chart for shielding as a function of time, distance, and amount of radioactivity for quantities of radium up to 100 g. This was based on a tolerance dose of 0.1 R/day which by this time had become well accepted nationally. In addition, this report made the first reference to precautions to be taken in the movement and transportation of radium within hospitals, between hospitals, and by means of public transportation.

RADIOACTIVE LUMINOUS COMPOUNDS (REPORT NO. 5)[59]

During the early 1930's considerable evidence of injury to radium dial painters began to develop from a large number of individuals who were severely or even fatally injured as the result of having ingested preparations containing radium and mesothorium. Because of this, various organizations associated with the dial painting industry requested the National Bureau of Standards to study the problem and this was taken up by the Advisory Committee in a special study.

A series of determinations was either verified or made anew relative to the question of radium injury. It was determined that the most serious injuries resulted when the compound was taken directly into the body either by mouth or by inhalation. At the same time it was demonstrated that when proper precautions were taken to prevent the compound from entering the mouth or lungs of the worker no detectable injury resulted.

It was determined that the maximum injurious effects were produced only after the radioactive material had entered the body and deposited in the bone. Therefore the radium had to first find its way into the bloodstream, and this could only occur in appreciable quantities through the mouth and lungs. A small fraction of each amount of radium taken into the body was subsequently deposited in the bone, and over a period of years a dangerous amount of radium could be accumulated in the skeleton as the result of the daily absorption of small amounts.

Since the accumulation is a slow process, even when appreciable amounts of the compound are ingested daily, it is not surprising that no immediate injury or illness appeared from the ingestion of the radioactive compounds even after a dangerous amount had been accumulated in the skeleton; the evidence is slow in developing. It was determined in the study that when deposits of radium are large the damage is chiefly to the skeleton, followed by damage to the white and red blood corpuscles resulting in leucopenia or anemia or both. It was shown that this occured with deposits of 12 to 100 micrograms of radium.

Unfortunately, the situation is such that radium, once deposited in the bone, remains there almost indefinitely and cannot be readily removed by simple medical treatment. Consequently, once the condition had been established there was little chance of restoring a normal condition in time to prevent serious effect. It was therefore deemed essential to avoid all ingestion and inhalation of radioactive luminous compounds and to test all workers periodically for exhaled radon in the breath. By this test small deposits of radium could be detected long before clincial symptoms had appeared. It was therefore recommended that any worker who showed a deposit of more than 0.1 microgram of radium, as revealed by the expired air test, change his occupation immediately and be treated by decalcification therapy or any other modality that had been developed for the purpose. It was determined that continued inhalation of radon or thoron could induce carcinoma of the lungs and also be a contributing factor in the anemia resulting from radium poisoning. It was recommended that the radon concentration in workrooms should not exceed 10^{-11} curies per liter of air.

The committee also examined the question of whole-body exposure of the worker to gamma radiation and reconfirmed the earlier tolerance level of 0.1 Roentgen for the working day.

In retrospect it is interesting, in view of today's knowledge, to look back to the recommendations of this particular report. Since that time hundreds of individuals having radium exposure records have been examined either in life or in post-mortem, and throughout all of this no important evidence has developed to indicate any urgent need to change the recommendations as to permissible body burden.

Having evaluated the nature of the problem and the process of radium poisoning, the committee then developed detailed recommendations for the handling of radioactive luminous compounds, including shielding, working procedures, test methods, and so on. Again, it is interesting to note that no urgent need for serious modification of these general precautionary recommendations has developed in spite of the vastly extended experience with radium over the last three decades. Cautions were also issued regarding the public transportation and shipment of the radioactive luminous compounds but it was not until some years later that extensive and detailed rules on transportation were developed.

Discussions up to this point cover fairly well the activities of the Advisory Committee on X-Ray and Radium Protection since its formation in 1929 up until the period just preceding the development of atomic energy, at which time a whole new era of radiation hazards and radiation protection needs was opened up. Before proceeding with developments in that era it may be desirable to return to some of the more important activities being carried out by international bodies.

INTERNATIONAL ACTIVITIES

The programs and activities of the International Committee on X-ray and Radium Protection during the early period had proceeded at a moderate level as noted above. It held its last meeting in 1937 and during the nine years of its existence had only a single change in membership.

In the late 1920's the Health Section of the Secretariat of the League of Nations responded to a recommendation made during its Twelfth Session, in which the health administrations in the various countries were invited to supply information on the regulations in force in this connection. The only known output of this action was the preparation of a document by Wintz and Rump outlining the situation up to about 1930.[18] Their careful and detailed report has provided one of the best compilations of early radiation protection activities and has already been referred to extensively in the discussions above. Apparently no official action was taken by the League and only a short time later the League of Nations fell into limbo.

Another activity contributing importantly to radiation protection were the programs of the International X-ray Units Committee. In 1928 this committee had agreed tentatively on the definition of the roentgen for x-ray measurement,[10] and in 1931 this was reviewed and confirmed.[194] With international acceptance of this unit there was a rapid consolidation of information and efforts leading to the conversion of dose measurements from skin erythema unit to Roentgens. This culminated in the adoption in 1934 of a tolerance dose by both the NCRP and the ICRP.

Because of the outbreak of war in Europe activities of both of the international committees came to a halt. So also did the activities of the Advisory Committee on X-ray and Radium Protection, after the issuance of its reports on the safe handling of radioactive luminous compounds in 1941 at about the time the United States was drawn into the conflict.

While there was some degree of continuity of organization of the two international commissions there was no further activity between 1937 and 1950. During this period, L. S. Taylor was asked to act as Secretary for the Protection Committee until it could again hold a meeting. When this time arrived, he and Dr. Rolf Sievert from Sweden were the only surviving members of the original committee.

In the middle 1930's the International X-ray Units Committee had been reorganized and Taylor was named the Secretary of this Committee. Here again there was a tragic change in membership, and of the 22 members meeting in Chicago in 1937 only 3 or 4 remained in 1950. We shall return to the reorganization and operation of these committees after recounting some of the activities which occurred during the 1940's.

MANHATTAN DISTRICT

During the early 1940's the major efforts in the radiation field were directed towards the development of controlled fission and nuclear weapons. That in itself is a separate story, but out of it came the generation of vastly expanded uses of all kinds of radiations and radioactive materials which in turn necessitated an increased understanding of control measures to a degree never dreamed of a few years earlier.

It was recognized from the outset of this atomic energy program that there might be severe radiation protection problems and consequently they organized to cope with whatever situations might arise. While many volumes (mostly classified) have been written on the radiation protection activities of this project, excellent summaries have been given by Dr. Robert S. Stone, who was the Director of the Health Division of the Metallurgical Laboratory in Chicago.[33, 34] This operation was immediately confronted with the necessity for making some kind of decision as to what amount of radiation might be acceptable by the project workers.

It was natural, in choosing a tolerance dose, that they start with the value proposed by the Advisory Committee in 1934 and in use in the United States at the time of the Plutonium Project. However, they quite properly felt that the tolerance dose levels had been arrived at on the basis of rather scant evidence and that every effort should be made to try to evaluate them on a more quantitative basis. As far as clinical observations were concerned there were little or no clear-cut demonstrations of what was described as permanent injury for doses of two or three or even more Roentgens per day for fairly extended periods. There was the general feeling that the tolerance dose of 0.1 to 0.5 R/day might not be unreasonable.

Relatively large-scale experiments were instigated at the National Cancer Institute, the University of Chicago, the University of Rochester, and Oak Ridge. At the Cancer Institute, mice, guinea pigs, and rabbits were exposed to gamma rays for 8 to 24 hours per day, 7 days a week, at dose rates ranging from 0.11 to 8.8 Roentgens per day for the lifetime of the animal. At Rochester, rats, rabbits, and dogs were exposed to 250 kV and 1000 kV x-rays for short periods daily - 6 days per week at dose rates ranging from 0.1 to 10 R/day. At Chicago and Oak Ridge, similar kinds of exposures were given to various animals to compare the effects of γ rays and neutrons. On the basis of observed skin damage, there was a surprising absence of effect, but there was obvious symptomatic evidence that the irradiated animals did not survive as along as the non-irradiated ones.[33]

Experiments with animals showed that gamma rays administered chronically, in the order of 10 R/day, produced no opacity of the cornea nor cataract formation, even for total doses as high as 12,000 R. On the other hand, acute doses as low as 400 R applied at the time of birth did cause eye lesions in mice. Corneal opacities were also produced in animals treated with beta rays delivered at the level of the order of 50 "rem" per day. As compared to this, very much lower doses of neutron radiation did produce cataracts.

In connection with the chronic irradiation of animals for other purposes, it was noticed that there was a significant reduction in the life span of animals exposed at doses of the order of 0.5

R/day. Inhibition of fertility was also studied in animals and it was found that doses in the order of 1 R/day produced little, if any, change in the average litter size. Doses of the order of 5 R/day, delivered for long periods of time, definitely impaired fertility.

The examples above are obviously inadequate today as measures of the adequacy of a tolerance or permissible dose. However, at the time they were carried out in the early 1940's, they lent comfort to the use of a tolerance dose level, equivalent to 1/10 R/day. As a matter of fact, their operations were all designed so as to maintain expsoures at levels substantially less and since that time there have been no individual cases, or a statistically significant number of cases, of identified injury to persons exposed during that trial period.

There was special concern, of course, for the possible genetic effects of x-rays upon the radiation workers in the plutonium project. By 1940 Muller's studies of genetic effects in drosophlia were universally accepted. Further experiments during the plutonium project were carried out with mice exposed at dose rates of the order of 0.1 to 10 R/day. Small effects were found for exposures of the order of 100 R. It was generally felt that there was no reason for great alarm to individuals as far as concerned genetic effects to exposures of 50 R at a rate of 0.1 R/day, but it was recognized that if a large portion of the population were to receive such exposures, the effects could be serious.

Thus it was, that because of national necessity, the crash program of producing vast amounts of radiation and radioactivity was undertaken with good confidence that unless very large numbers of people received relatively high exposures the hazard was not unacceptable. The many programs started during the war period have since been carried out and greatly expanded since then by the Atomic Energy Commission and collectively probably represent the major single source of radiation effects studies in the world.

OTHER PROTECTION ACTIVITIES (1940's)

Along with this intensive activity devoted to understanding the effects of radiation, and to shielding people from it, was the generation of a whole new discipline in science and technology frequently referred to as health physics. While the name itself was considered by some to be somewhat of a misnomer it generally includes multi-discipline training in radiation measurement and evaluation and the general assessment of health hazards due to radiation exposure.

Two other important but lesser activities took place in the early 1940's. One of these was recognition by industry for the need of better control of industrial radiation exposure. With the greatly accelerated used of radiation due to war activities, mainly in the examination and inspection of materials, large numbers of workers were being occupationally exposed to radiation in manufacturing plants, shipyards, and so on. Because of this, the American Standards Association organized its Committee Z-54 to undertake the development of suitable radiation protection codes.[185] This request had been made earlier to the NCRP, but at that time all of the NCRP members were heavily committed to other war efforts and were not able to add this to them. The ASA committee was set up under the sponsorship of the National Bureau of Standards and, in fact, was managed out of the same office as the NCRP and hence took advantage of information which had already been assembled for protection purposes. The ASA code was not completed until after the war but informally had come into extensive use since its drafts were widely circulated around the country. This report did not deal with the basic protection philosophy—they accepted that of the NCRP. Their main effort was to provide the many details of design and operation necessary for the safe use of x-rays in industry.

The other activity of importance was the development of elaborate and detailed regulations concerning the shipment of radioactive materials by public carrier. Need for this had been voiced by the NCRP in several earlier reports but when it was asked by the Interstate Commerce Commission to advise on the subject, the Advisory Committee members, as for the ASA problem, were already too heavily committed to take on additional tasks. Accordingly, this program was carried out under the auspices of a committee of the National Academy of Sciences and a first report and regulations were issued.

THE NATIONAL COMMITTEE ON RADIATION PROTECTION

By 1946 it had become evident that radiation uses in all categories would be rapidly expanded. Whereas in the 1930's the main concern had been for radiation workers, it was not clear that concern should be had for vastly larger numbers of workers and possibly relatively large population groups. During the preceding half decade the sustained and controlled nuclear reaction had been developed; radionuclides in great variety and in vast quantities could be, and were being, produced and were entering into research and medical channels. Accelerators such as the cyclotron which had been in limited use before the war had been further developed and their uses extended during the war period. Other high energy accelerators such as the Van der Graaf generators and betatrons, capable of producing energies up into the multimillion volt range, were now in being and their uses bound to expand.

In September 1946, a meeting of the Advisory Committee was called to discuss the extensive revision needed in the x-ray protection recommendations - particularly in the upper voltage region. At this meeting it was pointed out that protection problems had become too complex to permit their study and solution by a committee constituted as was the Advisory Committee at that time. It was recommended that steps be taken to secure participation of such additional groups as the Manhattan District, the U. S. Public Health Service, the military departments, etc. in an expanded work program. Accordingly, invitations were sent out to each of these agencies to appoint two representatives to attend a meeting to discuss future plans in the field of radiation protection.

Thus, the first post-war meeting of the committee was held on December 4, 1946.[35] The agenda of this meeting pointed out that new data had become available since the issuance of recommendations for x-ray and radium protection in the 1930's. Many new protection problems had arisen with the rapid expansion of activities in the radiation field, including protection against neutrons, multimillion volt x-rays, radioactive isotopes, and so on. It was suggested that the scope of the work be defined and that consideration be given to organizing small groups to deal with each of the problems and their completed reports to be submitted to the full committee for approval.

It was agreed that the Committee should be substantially enlarged and reorganized. At the same time, it was felt that the name of the Committee should be made more inclusive and it was therefore renamed National Committee on Radiation Protection (NCRP). The National Bureau of Standards was reaffirmed as the central coordinating agency for the work of the Committee; sponsorship by an impartial agency was felt to be particularly advantageous in view of the various types of participating organizations (radiological societies, industry, government, and the possible inclusion of industrial and labor groups).

The general organization and operational procedures outlined below were agreed upon at this meeting and became the basis for the continuing operation of the Committee:

(1) The committee would consist of an Executive Committee, Main Committee, and as many subcommittees as necessary to consider the problems that came within the Committee's scope.

(2) The Executive Committee would be composed of five members appointed by the chairman and subject to the approval of the Main Committee. The Committee chairman would act as chairman of the Executive Committee.

(3) The Main Committee would be composed of (a) technically qualified representatives appointed by organizations interested in the scientific and technical aspects of radiation protection, (b) representatives, at large, whose services were felt to be of special value, appointed by the Executive Committee, and (c) chairmen of subcommittees.

(4) The choice of chairmen and members of subcommittees would not be restricted to members of the Main Committee but would be based on the particular qualifications needed for the work. (In organizing subcommittee memberships, the following practice was followed: The subcommittee chairman was selected by the Committee chairman with the approval of the Executive Committee; the subcommittee chairman chose his working group, made informal contacts, and submitted to the Committee chairman his membership selections; the chairman issued formal membership invitations to serve on the subcommittees. Additions could be made to the subcommittee membership if particular specialized information was found to be needed, or indivi-

duals could be invited to serve as consultants to the group with due acknowledgment of their assistance included in the published recommendations.)

(5) The final reports of the subcommittee were submitted to the Executive and Main Committees for approval. Because of the high degree of success of the NBS Handbook series, it was recommended that this mode of publication and distribution to the public be continued.

Because of the reorganization and enlargement of the Committee, the chairmanship was thrown open for reconsideration. L. S. Taylor was nominated and approved by vote to continue indefinitely in this capacity.

Discussions were held regarding the organizations that might appropriately be invited to participate and the suggestions made were used as a basis for the subsequent enlargement of the representation on the Main Committee. It was agreed to establish the following subcommittees:

1. Permissible external dose.
2. Permissible internal dose.
3. X-rays up to 2 MeV.
4. Heavy ionizing particles (neutrons, protons, and heavier).
5. Electrons, radium, and γ rays above 2 MeV.
6. Radioactive isotopes, fission products, including their handling and disposal.
7. Monitoring methods and instruments.

All seven committees began their assignments immediately and within a year drafts of reports had been prepared by all of them. In the meantime, in addition to the original group meeting in 1946, the main committee was expanded to include representatives of the several radiological organizations so that the main committee consisted of some 26 individuals supported by a substantially larger number of individuals working with the various subcommittees.

NCRP REPORT ON EXTERNAL EXPOSURE[7,1]

Although all of the subcommittees had assignments to prepare reports covering their particular field of interest, perhaps the most basic subcommittee, and the one upon which all the others depended, was that of Subcommittee 1 under the chairmanship of Dr. Failla. Even the permissible doses for radioactive material within the body were related back to the basic information on the effect of external radiation in the form of moderate energy x-rays or gamma rays. All codes of practice and other recommendations relative to the safe handling of radioactive materials went back to the basic concept of permissible dose which was to be developed by Sub-committee 1.

For various reasons, the formal issuance of the report of this subcommittee was delayed for a considerable time even though the substance of its recommendations had been fed into the other subcommittees and their recommendations. No small item in the preparation of this report were the discussion and explanations that were thought to be essential to the complete presentation of the subject. Up until this time none of the reports from any of the national or international protection committees had included a discussion of the reasons why they made their particular recommendations. Therefore, it seems appropriate that a brief outline of the historical background of the work of this subcommittee be presented.

Following a considerable amount of preliminary correspondence, informal meetings, and orientation, the first formal meeting was held in Chicago in June 1948. At this time a number of basic and far-reaching decisions were made: 1. to lower the then-existing permissible dose of 0.1 R/day by a factor of about 2 and to express it in terms of a week; 2. to take the blood forming organs as the most critical tissue and to apply the permissible limit of 0.3 R/wk to these organs; 3. to consider skin as a critical organ and to set the permissible dose for it at 0.6 R/wk at a depth of 7 mg per square centimeter; 4. to allow a permissible dose for persons over 45 years of age double that for younger adults; 5. to allow larger weekly doses for the hands and feet (1.5 R/wk); 6. to make suitable recommendations about accidental exposures involving a single dose that might be as large as 25 R; and 7. to recommend an RBE (Relative Biological Effectiveness) of 1 for x-rays, gamma rays, and beta rays, 5 for thermal neutrons, 10 for fast neutrons, and 10 for alpha particles.

In September 1948, the Chairman, Dr. Failla, attended some conferences in England, at one of which he presented a report on permissible exposures to ionizing radiations outlining the above points upon which the NCRP had agreed. Informal, but extensive, discussions on the subject of permissible dose were held with some members of the British counterparts of our committee and these led to informal acceptance of mutually agreeable permissible exposure values. Agreement was also reached on the concept of the critical organs for which permissible doses would be specified.

A preliminary report by this committee was completed by the summer of 1949 and, except for details, did not differ materially from the report finally published in 1954.[71]

In the meantime, a large number of copies of the report were circulated to various interested groups in this country and abroad so as to benefit by their comments and suggestions. As it turned out, most of these were in the nature of detail and there seemed to be general acceptance of the seven principles outline above.

In the fall of 1949, a series of conferences was undertaken between representatives of the United States, England, and Canada for the purpose of discussing radiation safety problems, with particular reference to atomic energy operations. The first of these "Tripartite Conferences" was held in Chalk River in 1949, at which time the more essential recommendations of the Failla report were accepted for exposure to external radiation. One change of importance made at these meetings was the raising of the RBE for alpha particles to a value of 20. The permissible exposures to internal radiation and the maximum allowable radioactive contents of air and water, initially adopted at the Chalk River Conference, were related back to the basic philosophy of permissible exposure developed by the subcommittee on external exposure and the subcommittee on internal dose under the chairmanship of Dr. K. Z. Morgan.

At the meeting of the International Commission on Radiological Protection (ICRP) held in London in the summer of 1950 the essentials of the NCRP and Chalk River recommendations were adopted.[103] Immediately after the meetings of the ICRP in London, a second Tripartite Conference was held in Harwell, England, at which only minor changes were made in some of the permissible radiation exposures for internal emitters.

The next full meeting of the subcommittee was held in Chicago in December 1949, at which all of the ideas incorporated in its first report were discussed in detail and were adopted in principle. While there were no basic changes in principles since the first report in 1949, considerable time was spent on the wording of some of the recommendations and especially on the definition of permissible dose. The last meeting of the committee was held in Cincinnati in December 1952 at which time the principal discussion was related to genetic effects. Considerable discussion centered about the per-capita dose for the whole population during its reproductive age. As this was clearly a controversial subject and one involving a great deal of basic philosophy in addition to basic science, it had to be treated with great care and a number of geneticists were brought together for the purpose of assuring their agreement.

As already noted, this first report of Committee 1 discussed for the first time the basic biological factors and philosophical approaches that led up to the committee's recommendations.

As a matter of committee policy any controversial questions were debated and argued out until agreement was reached; when agreement could not be reached the item at issue was omitted.

The third Tripartite Conference was held in Harriman, New York, in March 1953, but no changes in basic philosophy developed at this conference. Later in 1953, at a meeting at the International Commission on Radiological Protection in Copenhagen, a report of this subcommittee, with minor deletions of topics not of international interest, was used as a basis for their discussions. The general principles developed were followed closely by the ICRP.

It is worth noting that in the 20 years since the report of subcommittee 1 was prepared the subject has been studied further in vast detail yet there have been no important or fundamental changes in the philosophical approach to the problem. There have been changes in numerical values, and there have been some developments, particularly in the genetic field, which have modified the approach to the genetics problem, but these have not seriously changed the original recommendation.

In its section under Radiobiological Considerations, various factors influencing radiation protection standards were discussed. These included the biological variability of individuals and organs. There was also discussion of the problems introduced by the long latent periods that might exist between radiation exposure and the time that injury or effect might become evident. Questions of recovery of biological change and repair of injury together with the time factors involved in radiation exposure were included. Attention was directed to the radio sensitivity of different organisms and organs for comparisons of radiation effects due to different kinds of radiations—for example, neutrons or alpha rays as against x- or gamma rays—leading to the development of the concept of relative biological effectiveness. The subject of differential variations involving effects of different radiations on different organs and the problems of distinction between whole body irradiation versus partial body irradiation were discussed. Also, there was considerable discussion of genetic effects and the effects of radiation on life span.

In the section under Protection Criteria the general concept of acceptable risk was discussed and the philosophical approach of risk in relationship to cost or benefit was developed for the first time. Then, after having defined or discussed the question of critical tissues, outlines and discussions of permissible dose, permissible weekly dose, and maximum permissible dose were developed. The problem of individual organ dose was developed especially for application by Subcommittee 2 to internal emitters.

In the section on permissible weekly doses for critical organs the problem of long-range exposure to x-rays was developed with special emphasis on exposure under the conditions of radiological practice. Also, long-term exposure of the blood forming organs, skin, gonads, lens of the eye, and other organs was outlined.

The permissible dose was defined as "the dose of ionizing radiation that, in the light of present knowledge, is not expected to cause appreciable bodily injury to a person at any time during his lifetime. As used here 'appreciable bodily injury' means any bodily injury or effect that the average person would regard as being objectionable and/or competent medical authorities would regard as being deleterious to the health and well being of the individual."

"For the skin the proper value is obviously the highest dose received by any skin area (of the order of a square centimeter). For the bloodforming organs an average value of some sort is required. An average dose for a significant volume of the bloodforming organs in the body region in which the tissue dose is highest is satisfactory. The 'significant volume' may be assumed to be of the order of 1 cm. For the gonads the pertinent dose may be assumed to be the average dose in a significant volume of the organ in the region of highest tissue dose. The significant volume in this case may be taken to be 10% of the total volume of both gonads. For the lens, which is very small, the significant volume is the volume of the lens of either eye."

Following this was further discussion of long-term exposure related to different kinds of radiation - especially for radiations having widely different RBE's and mixed radiations. Comparison of exposures due to external and internal sources as influencing long-time effects was discussed in detail.

It was recognized that when considering long-term exposure of the type that might be involved by radiation workers, or even the population in the event of widespread radiation usage, there might be modifying factors involving age, weekly dose fluctuation, influence of non-occupational exposures, and the numbers of exposed individuals.

Then, following the development of the background material, as outlined very briefly above, a series of protection rules was developed. While the numerical values involved in these rules have since been modified, the basic principles that were involved in the rules are still generally applied:

LONG-TERM EXPOSURE
(quoted from NCRP Report No. 17[71])

"**Rule I. Ionizing Radiation of Any Type or Types**

For adults under 45 years of age whose entire body, or major portion thereof, is exposed to ionizing radiation from external sources for an indefinite period of years, the maximum permissible total weekly doses shall be 300 mrems in the bloodforming organs, the gonads, and the lenses of the eyes; 600 mrems in the skin; and the respective values of the weekly doses in millirems in all other organs and tissues of the body according to the basic permissible dose distribution. For persons 45 years of age or older similarly exposed, the corresponding maximum permissible total weekly doses shall be double the above stated values, provided that the portion of the weekly dose in the lenses of the eyes contributed by radiation of high specific ionization does not exceed 300 mrems.

Rule II. X-rays (Roentgen Rays, Gamma Rays) with Photon Energy Less Than 3 MeV

For adults under 45 years of age whose entire body, or major portion thereof, is exposed solely to x-rays with photon energies less than 3 MeV from external sources for an indefinite period of years; the maximum permissible total weekly dose shall be 300 mR measured in air at the point of highest weekly dose in the region occupied by the person, provided that the actual total weekly dose in the gonads does not exceed 300 mrads. For persons 45 years of age or older similarly exposed, the corresponding maximum permissible total weekly doses shall be double the above stated values, provided that the actual total weekly dose in the lenses of the eyes does not exceed 600 mrads.

Rule III. Radiation of Very Low Penetrating Power (Half-Value Layer Less than 1 mm of Soft Tissue)

For adults of any age whose entire body, or major portion thereof, is exposed to ionizing radiation of very low penetrating power from external sources for an indefinite period of years, the maximum permissible total weekly dose in the skin shall be 1500 mrems, provided that the total weekly dose in the lenses of the eyes does not exceed 300 mrems.

Rule IV-A. Local Exposure of the Hands and Forearms to Any Ionizing Radiation

For adults of any age whose hands and forearms are exposed to ionizing radiation from external sources for an indefinite period of years, the maximum permissible total weekly dose shall be 1500 mrems in the skin, provided the respective weekly doses in milirems in all other tissues of the hands and forearms are not in excess of those that would result from exposure to ordinary x rays at a weekly dose of 1500 mr in the skin.

Rule IV-AX. Local Exposure of the Hands and Forearms to X-Rays (Roentgen Rays, Gamma Rays) of Any Photon Energy

For adults of any age whose hands and forearms are exposed solely to x-rays from external sources for an indefinite period of years, the maximum permissible total weekly dose shall be 1500 mr in the skin.

Rule IV-B. Local Exposure of the Feet and Ankles to Any Ionizing Radiation

For adults of any age whose feet and ankles are exposed to ionizing radiation from external sources for an indefinite period of years, the maximum permissible total weekly dose shall be 1500 mrems in the skin, provided the respective weekly dose in millirems in all other tissues of the feet and ankles are not in excess of those that would result from exposure to ordinary x-rays at a weekly dose of 1500 mR in the skin.

Rule IV-BX. Local Exposure of the Feet and Ankles to X-rays (Roentgen Rays, Gamma Rays) of Any Photon Energy

For adults of any age whose feet and ankles are exposed solely to x-rays from external sources for an indefinite period of years, the maximum permissible total weekly dose shall be 1500 mr in the skin.

Rule IV-C. Local Exposure of the Head and Neck to Any Ionizing Radiation

For adults whose heads and necks are exposed to ionizing radiation from external sources for an indefinite period of years, the maximum permissible total weekly doses shall be 1500 mrems in the skin and 300 mrems in the lenses of the eyes, provided the respective weekly doses in milirems in all other tissues of the head and neck are not in excess of those that would result from exposure to ordinary x-rays at a weekly dose of 1500 mr in the skin. For persons 45 years or older the weekly dose in the lenses of the eyes may be 600 mrems, provided that the portion contributed by radiation of high specific ionization does not exceed 300 mrems.

Rule IV-CX. Local Exposure of the Head and Neck to X-rays (Roentgen Rays, Gamma Rays) of Any Photon Energy

For adults whose heads and necks are exposed solely to x-rays from external sources for an indefinite period of years, the maximum permissible total weekly doses shall be 1500 mR in the skin and (a) 450 mR in the lenses of the eyes of persons under 45 years of age, (b) 600 mr in the lenses of the eyes of persons 45 years of age or older.

OCCASIONAL EXPOSURE

Rule V-A. Accidental or Emergency Exposure to X-rays (Roentgen Rays, Gamma Rays) with Photon Energy Less Than 3 MeV

Accidental or emergency exposure of the whole body of adults or parts thereof to x-rays with photon energy less than 3 MeV, from external sources, occurring only once in the lifetime of the person, under the conditions and in the respective dosages stated below, shall be assumed to have no effect on the radiation tolerance status of that person.

(a) Exposure of the whole body - any adult. Total dose, measured in air: up to 25 R.

(b) Local exposure - any adult. Dose measured in air and additional to whole-body dose: (1) Hands and forearms, up to 100 R; (2) feet and ankles, up to 100 R.

Rule V-B. Planned Emergency Exposure

Emergency work involving high-level exposure to x-rays with photon energies less than 3 MeV shall be carried out on the basis that the person will not receive doses higher than one half the respective doses stipulated in Rule V-A. If the doses actually received in the performance of such work do not exceed the respective maximum doses stipulated in Rule V-A, the exposure may be considered to be in the category covered by Rule V-A. Women of reproductive age shall not be subjected to planned emergency exposure.

Rule V-C. Accidental or Emergency Exposure to Other Types of Ionizing Radiation

Rules V-A and V-B are applicable to accidental or emergency exposure to ionizing radiation of any type and energy when the tissue doses resulting therefrom in the different organs and tissues of the body (expressed in rems) do not exceed numerically the respective tissue doses in rads resulting from exposure to x-rays with photon energy less than 3 MeV, under the conditions stipulated in Rule V-A, provided, however, that the portions of the respective tissue doses in rems contributed by radiation of high specific ionization do not exceed 50% of the total tissue doses.

In view of the breadth of the subject that had to be studied and evaluated in arriving at basic protection philosophy, it is perhaps understandable that nearly five years elapsed between the time that the basic aspects of the report were completed and the time that it actually appeared in print. Were it not for the fact that the contents were made known on a wide base and were used by the NCRP as well as by organizations in other countries, serious confusion could have resulted through the introduction of a multiplicity of protection concepts and principles. This appears to have been substantially avoided."

NCRP REPORT ON INTERNAL EMITTERS

Committee 2, on "Permissible Internal Dose", under the Chairmanship of Dr. Karl Z. Morgan, undertook the task of dealing with the many parameters necessary for the establishment of maximum permissible amounts of radioisotopes in the human body and maximum permissible concentrations in air and water. The report, entitled Maximum Permissible Amounts of Radioisotopes in the Human Body and Maximum Permissible Concentration in Air and Water, was published as NCRP Report No. 11 in March 1953.[65] While a considerable body of information had been developed during the Manhattan District operations it was necessary to study these preliminary results further and to substantially extend the concepts to many other radionuclides that were already coming into experimental use.

One of the greatest difficulties encountered in the preparation of the report lay in the interpretation of the existing biological data dealing with the uptake and retention of radioactive material by the body. Many variables were present in each experiment and major discrepancies occurred even between the most reliable researches. A tremendous effort was necessary to obtain a better understanding of the biological effects of radiation and, as a matter of fact, heavy efforts are still required along this line. In the three or four years during which time this report was in preparation, progress was being made so rapidly in the field that at times it seemed almost hopeless to keep abreast of the changes. Nevertheless, the numerical values given in their first report were such that it was believed that errors - if any - would be in the direction of providing additional safety. It was generally agreed that though often impractical, if not impossible, to prevent some radionuclides from entering the body, it was nevertheless necessary to avoid unnecessary exposure to radionuclides as far as possible. It was considered desirable, therefore, to establish levels of maximum permissible exposures to serve as guides to safe operations and to upper levels of exposure.

In some cases there was considerable uncertainty about the maximum permissible values given in the initial study, but because many persons were already being exposed to certain of the radionuclides it was considered essential to agree upon what might be considered as safe working levels for those nuclides rather than wait until more complete information was available. In this connection it had to be borne in mind that persons might be exposed to radionuclides for an indefinite period of time, possibly a lifetime.

Because in general it was impossible to predict at the start how long a person might be exposed, permissible limits had to be set on the assumption that occupational exposure would continue throughout the working lifetime of the individual and that environmental exposures could continue for a full lifetime. Thus, the values given in the report were derived on the basis of continuous lifetime exposure or for equilibrium conditions in which the rate of elimination had become equal to the rate of deposition in the body. Events have proven that the original recommendations have for the most part been reasonably conservative, although there have been some important changes necessary as a result of new information of different exposure conditions.

In its final results the committee presented maximum permissible amounts of radioisotopes allowable in the total body and maximum permissible concentrations of air and water for continuous exposure. The listing covered some 75 radionuclides which were then considered to be the most important ones in view of their utilization in research and medicine. In addition to the end results, the report included tabular information on the constants used in carrying out the calculations for each radionuclide. Since many of the end results could not be obtained or evaluated from direct experiment, calculational methods had to be developed for a great majority of them. To this end the committee set up models and appropriate equations to cover the conditions as best they could be evaluated at the time. Much of this formulation is still valid although some of it has been refined to meet either special conditions or new knowledge subsequently developed. It might also be mentioned that some of the uncertainties recognized in the earlier report, prepared around 1949, are still not cleared up satisfactorily as late as 1969, two decades later. This is not an indication of lack of progress but does point out the enormous difficulty of arriving at reliable and accepted end results in the absence of pertinent experimental or clinical observation.

In the process of developing the background information and models for the work, the committee had to make comparisons between incidence of injury caused by radiations from artificially produced radionuclides as compared with those from x-rays or gamma rays from radium on which there was considerable experience prior to 1940. It was not unnatural that these comparisons would play a critical role in the evaluation of effects from new radionuclides whose metabolic behavior in the body might be basically quite different from that of naturally occurring radioactive material. Nevertheless, this has continued to serve as a sound base even though there is an increasing amount of biological information now available as a result of carefully planned animal experiments with artificial radionuclides.

In carrying out these comparisons it was necessary to critically evaluate not only the experiments made with animals but the clinical experience obtained with man; there is a moderate degree of consistency between the two. On the other hand events have also shown that there is some risk in trying to make simple extrapolations from one to the other.

It was also considered necessary to make comparisons between the exposure from naturally occurring radiation in the background and that from the radionuclides in our bodies and in the air we breathe and in the water and food that we consume. Whether or not such radiations are regarded as necessarily being deleterious, it was recognized from the outset that a considerable number of factors would determine the potential hazard of the radionuclide once it has been taken into the body system. For example, the effects would certainly depend on the quantities of nuclide available from outside the body, and which by one means or another could enter into it. Then, once the radioactive material had entered the body it was important to understand the mechanism of its retention and elimination. As a part of this it was necessary to develop a knowledge, either from experiment or claculation, about the fractions that would go from the blood to the critical body tissues.

Of critical interest, of course, was the understanding of the radio sensitivity of the tissue and this is a factor that could only be obtained by some type of observation.

Many of the radionuclides would selectively deposit in certain organs. Estimates of the quantities deposited could be obtained by means of studies made with non-radioactive nuclides of the same element as, for example, normal calcium versus radioactive calcium. Important in the overall end result would be the size of the critical organ. A given amount of a certain radionuclide deposited in an organ of, say, two or three grams might have a more serious effect on that organ than the same amount of radioactive material deposited in an organ of two or three hundred or two or three thousand grams. Important also was some kind of evaluation of the essentiality of a critical organ to the proper function of the whole body and the extent to which deposits in different organs might add or even have synergistic effects.

An especially difficult problem relative to a particular organ would be the dosimetry involved in evaluating the effects of several different kinds of radionuclides that might deposit in the organ and contribute to the ultimate damage. Various pragmatic schemes were developed for making this evaluation and again time has shown that the early ones were generally good.

Important in the overall picture of internal emitters is the matter of biological half-life which is a combination of physical and radiological half-lives of radionuclides combined with factors of concentration and ultimate elimination. An essential part of this was a good knowledge about the energy of the radiation released to the tissue by the radionuclides. For this it was necessary to have good knowledge of the decay schemes for each radionuclide and extensive studies of this question are still in progress.

NCRP REPORT ON X-RAY PROTECTION

Committee III under the Chairmanship of Dr. Harold O. Wykoff on "Protection against X-rays up to Two MEV" undertook as its primary objective the development of what might be regarded as a code of practice primarily directed towards the medical uses of x rays. Nevertheless, the general principles and much of the material developed by this committee was equally useful in x-ray operations in industry and have since been utilized by various industrial standards groups. Since this committee had the earlier recommendations of the Advisory Committee on X-ray and Radium Protection to fall back upon, it was not surprising that they could move with greater speed than the other committees that for the most part were having to break new ground.

The committee began its activities in 1946 and its first full report was published in March 1949.[60] It is worth noting that since the new basic permissible exposure level for radiation workers had been set at 0.3 rem per week, its first announcement was in the report of Committee III (p.iii).

Before proceeding with the development of recommendations for working rules, practice, and radiation shielding, this committee laid down certain ground rules and recommendations relative to the use of radiation. It was pointed out that while the recommendations were intended primarily for protection of the radiation worker, nevertheless the report did include the more important elements of patient protection and the philosophy of limitation of the dose to the patient to that necessary for the medical purpose, and the avoidance of unnecessary or unproductive exposure.

It was emphasized that the increasing use of radiation and of radioactive materials in medical procedures made it increasingly necessary for the medical profession to exercise great caution and restraint in the use of x-rays for diagnosis and treatment of non-malignant diseases. It was emphasized that current methods of practice should be reviewed to see whether the same results could be obtained with less radiation or even by other means not involving radiation. In particular, it was emphasized that the beam of radiation reaching the patient should be as small in cross section as is essential for the examination, especially in fluoroscopy. It was also emphasized that the gonads should be protected whenever practicable, either by limiting the size of the beam or by local shielding.

Emphasis was placed on the necessity for dermatologists and radiologists to avoid as much as possible the use of radiation in the treatment of benign conditions of the skin, such as for fungus infections. This was considered to be especially important in the case of people handling radioactive material or being otherwise occupationally exposed to radiation. It was felt to be the duty of the physician to ascertain whether the patient had been occupationally exposed to radiation before prescribing radiation treatment and, if necessary, to consider the combined effects. It should be the duty of the patient to tell the physician when he is occupationally exposed to radiation whenever the physician prescribes radiation treatment. As a matter of fact, if a patient's condition is serious enough to warrant either treatment or elaborate diagnostic radiation techniques, it is usually unlikely that these should be avoided or postponed because of possible combined effects of the two kinds of exposure. Events have proven that normal occupational exposures at the new permissible levels are generally unimportant in reaching decisions as to medical procedures. They should, however, not be overlooked.

For consistency in presentation as well as ease of application, radiation sources were divided into seven categories and a set of both operating and shielding rules was given for each. These included: 1. fluoroscopic installations, 2. radiographic installations, 3. dental applications, 4. special requirements for mobile diagnostic equipment, 5. fluorographic equipment, 6. therapeutic installations operating at potentials up to 2250 kV, and 7. therapeutic installations for use up to 2000 kV.

A section on Electrical Protection was continued as a carry-over from the earlier days since, in spite of the introduction of shock-proof cables and fully shielded equipment, it was still possible for electrical accidents to occur. Electrical protection recommendations were especially important for the use of x-ray equipment in operating rooms or in the fighting of fire in areas where high voltage might exist.

An extensive appendix was given, including radiation attenuation curves for all of the voltages discussed in the report. In addition, based on these, it gave the approximate lead thicknesses required to reduce the radiation to given fractions of the useful beam. All of these calculations were based on the new permissible dose level of 0.3 rem per week and it was pointed out that should basic permissible occupational exposure levels be changed in the future these same tables could be readily corrected to meet the new needs. The appendix also included tables for the ready calculation of protection afforded by distance from given sources. Also, distinction was made between primary and secondary protective barrier requirements which in many instances, if applied literally, could result in substantial savings in the construction costs of x-ray installation.

NCRP REPORTS ON HANDLING AND DISPOSAL OF RADIOACTIVE MATERIALS

The development of the protective recommendations by the several other committees was started in the late 1940's but was under varying pressures and in widely different frameworks of available information; hence, the reports and recommendations appeared at irregular intervals. Work will be outlined in more or less chronological order covering the period up to about 1956 at which point there was an important shift in our basic radiation protection standard.

In the original list of committees, one on "The Handling of Radioactive Isotopes and Fission Products" was established but was soon divided up into several other committees when it was found that the workload was too great and the nature of the problems too diverse. However, the original committee under the Chairmanship of H. M. Parker proceeded immediately to develop a series of recommendations on the broad problem of "Safe Handling of Radioactive Isotopes" and its first report was issued in September 1949.[61] This project involved an entirely new line of activity because, with the exception of work in the radium industry and some low-level work in research laboratories, information on the problems facing radiation workers utilizing a large range of radionuclides was not yet available to the public. However, some members of the subcommittee had had an enormous amount of experience with the Manhattan District to draw from, and this was utilized to a major extent by the subcommittee.

Because this was such a new field, when the committee had completed its initial report, several hundred draft copies were circulated to all leading workers and authorities in the field for comment and criticism. As a result, the final report embodied all the pertinent suggestions received from these people and as the sciences have developed, the broad recommendations of this committee are still regarded as fundamental.

The first problem facing the subcommittee was to consider the most commonly available radionuclides and to arrange them in some order of importance which would take into account the quantities of material being used and its hazard potential in terms of uptake and concentration within the body. Effective radio toxicity was obtained from a weighting of the following factors for each radionuclide: A. half-life, B. energy and character of the radiations, C. degree of selective localization in the body, D. rates of elimination, and E. quantities involved and modes of handling typical experiments.

As a result of the initial study of these problems, it was possible to group the radio-nuclides according to their relative toxicity and in three broad categories considered as "low", "intermediate" or "high-level" in laboratory practice.

It was immediately clear that there were no sharp dividing lines as one moved from "low" to "intermediate" to "high-level" quantities of activity. Neither was it possible to draw any sharp dividing line between hazards in regard to being slight, moderately dangerous, or very dangerous. Nevertheless, it was possible to arrange a structure and logical grouping which turned out to be of considerable importance in planning radiation activities and in designing experiments. This general type of grouping, with only minor changes, is still in use today.

It was also possible to arrange the known hazards in some order of importance as follows: 1. exposure of the hands or other limited parts of the body to beta or gamma radiation, 2. exposure of the whole body to gamma radiation, 3. exposure of the whole body to beta radiation, and 4. deposition of radioisotopes in the body.

The fundamental principles underlying protective measures were: 1. to prevent ingestion, inhalation, interstitial, or other modes of entry into the body, and 2. to reduce the amounts of external radiation to permissible levels.

A second section dealt with the problems of personnel, including their selection and instruction and their medical and physical examinations, and finally took up the details of personal cleanliness, housekeeping, and supervision.

The third section dealt with the basic features of radioactivity laboratory design and equipment covering general working conditions, room construction, ventilation equipment, hoods and benches, and finally the disposal of contaminated wastes and the use of protective clothing.

A fourth section dealt with the required instrumentation for the proper evaluation of radiation exposure. Here the problem was divided into two main parts, that of personnel monitoring and that of area monitoring. During this early period, while there were pocket ionizations, chambers, film badges, etc. available, the concepts were still in somewhat rudimentary stages and relatively crude by today's standards. Nevertheless, it was possible to provide recommendations as to the basic requirements and accuracies of such instrumentation. Finally; the report dealt with details of hazard monitoring, itself. This covered inspection of personnel in work areas, protective clothing, wastes, and instrumentation. It also dealt with recommendations covering the management of radiation accidents for various types of exposure, including external, internal, surface contamination, minor injuries, inhalation, etc.

As already noted, a committee of the National Research Council has been studying the problems of interstate shipment of radioactive material. The report included reference to the first comprehensive requirements of the Interstate Commerce Commission, the Post Office Department, and the Interim Regulations for Air Shipment as they had been established by the end of 1948. In addition, the report included a section covering the movements of radioactive materials in the laboratory and provided data and information for beta ray and gamma ray shielding.

NCRP REPORTS NOS. 8, 9, 12, AND 16

A broad category of problems dealing with the laboratory wastes and their disposal for the relatively small laboratory was subdivided and managed by Dr. James H. Jensen. This committee completed three reports in short order and started two more which really never bore fruit.

The first of its reports (No. 8) was titled, The Control and Removal of Radioactive Contamination in Laboratories.[62] Although much of the ground work had been started by Committee II dealing with Internal Emitters, one of the greatest difficulties encountered in the preparation of the reports by this committee lay in the general uncertainty regarding permissible radiation exposure levels for ingested radioactive materials. Nevertheless, as it turned out, the recommendations adopted some years later did not differ sufficiently from those which were available to the committee at the time, with the result that their report was based on information which has remained relatively firm for the past two decades.

One of its important recommendations, and one that has now become standard practice, was that there should be established for each institution working with radioactive materials, an Isotopes Committee, whose responsibility should be the surveillance of all activities involved in the handling and use of radioactive isotopes. It was recommended that each such committee should designate one of its competent members to act as the responsible radiological safety officer or health physicist for all operations involving radionuclides. It would be his responsibility to review and approve working areas and protective facilities and to designate precautionary and safety measures for unusual situations. He also had to maintain inspection and monitoring procedures and to keep supervisors informed of all radiation exposure hazards and other factors affecting permissible working times in the given area. He should aid in the solution of new protection problems as they arose and direct that proper persons be notified in the event of emergencies, such as spills, injuries, fires, etc. He also should be charged with the arrangement and maintenance of adequate records of operations with radioactive materials and each individual's exposure. Furthermore, it was recommended that the Isotope Committee designate personnel who shall have the entire responsibility for the receipt, handling, labeling, storage, issue, disposal, and records of all radioactive materials received and shipped by the institution. The general principles of these recommendations have now become standard and accepted practice.

The committee encountered considerable difficulty in coming to grips with the problem of contamination. It was, of course, straightforward to recommend that contamination be avoided, that means be provided for minimizing it when it could not be completely avoided, and when it did occur that it be removed as promptly as possible. The big problem arose in trying to find some basis for specifying when it was adequately removed and provide a test for it. It was obvious from the outset that removal of contamination under all circumstances is rarely possible. The question is how does one specify what is permissible to leave behind? How clean is clean, and how do you measure it?

Various decontamination procedures were outlined with special attention being given to the removal of contamination from the body. The procedures suggested or recommended still remain valid under most circumstances but there have been definite improvements, especially in the application of the detergents. Special concern was expressed for the problem of dealing with wounds through which contamination might enter the body. It was generally agreed that rather than try to establish set procedures, each case should be treated individually and on its own merits by competent medical authority.

Decontamination procedures were also outlined for hospital use involving clothing and bedding and the cleaning of floors.

Control of contamination in laboratories was dealt with as a special case because of the large quantities of material that might be present during any particular procedure.

A second report (No. 9) prepared by the committee, operating under J. H. Jensen, developed recommendations on "Waste Disposal of Phosphorus-32 and Iodine-131 for Medical Users."[63] Their report was issued in November 1951.

In this early period, two of the major uses of reactor-produced radionuclides were in the therapeutic and diagnostic applications of iodine-131 and phosphorus-32. Individual quantities in use were not large but the sum total over a month's time might be fairly substantial. In addition, the uses were likely to be scattered widely in any urban area because both of these radionuclides had relatively short half-lives. It was deemed possible to hold the waste material in storage for some reasonable period during which the radioactivity could decay and after which the material might be disposed of by dumping into the normal sewerage system.

After some study, this type of practice was evaluated and considered to be satisfactory provided certain limitations were met and the amounts discharged within any one sewer connection controlled. The report dealt with questions of how to treat the material being held for decay, how to release it into the sewerage system in a gradual process, and in what amounts it might be reasonably released. A table was given for the number of millicuries that might be disposed of in any single event. This was given in terms of the numbers of beds in a hospital or separately in terms of the numbers of people involved. Recommendations were also given for individual disposal in small hospitals or in homes where patients might be assigned during the course of the treatment. An appendix to this report gave data and calculational methods necessary to deal with special cases of disposal.

A third report (No. 12) was issued by Dr. Jensen's committee providing recommendations for "The Disposal of Carbon-14 Waste."[66] This was issued in October 1953. As in the case of iodine and phosphorus, this report was prepared for the rapidly expanding experimental and medical uses of Carbon-14 where the individual quantities of material to be disposed of would rarely be very large but where they might be relatively frequent.

A number of disposal methods were considered: (1) isotopic dilution in a ratio of at least one microcurie to ten grams of stable carbon, (2) sewage disposal at levels not exceeding one millicurie per hundred gallons of sewage; (3) incineration at levels not to exceed five microcuries of ^{14}C per pound of combustible material, (4) atmospheric dilution at a rate not to exceed 100 microcuries per hour, per square foot of intake area of the hood, (5) garbage disposal at a rate not to exceed one microcurie per pound of garbage, (6) burial, when the concentration levels did not exceed five microcuries per gram and when the total amounts mixed with a cubic foot of soil did not exceed ten millicuries.

In addition to recommending such details of disposal as noted above, the committee also worked out the permissible dose in relation to Carbon-14 disposal and gave procedures for calculating the exposure levels under varied conditions.

Early plans of this committee had been to develop similar recommendations for other more commonly used radionuclides, but in the meantime the Atomic Energy Commission's disposal programs had begun to develop and it was felt that the problem was not of the same urgency as it had been a few years previously.

However, one problem remained in connection with the disposal of large amounts of radioactive material. The committee under Dr. Jensen examined, at some depth, the problem of disposal of radioactive materials by large-scale burial in soil or by incineration. Neither of these reports was completed. This was due in part to the great complexity of the problem and partly because in the meantime it was clear that federal laws were going to be developed which would put this responsibility in the hands of the Atomic Energy Commission. The study of incineration has continued on an irregular basis for a number of years and it is still doubtful at the time of this writing as to whether a satisfactory public disposal means by incineration can be evolved.

In the meantime it became obvious that large quantities, both in bulk and activity, were going to be disposed of in the ocean; therefore, the fourth report by Dr. Jensen's committee (No. 16) devoted its attention to the question of "Radioactive Waste Disposal in the Ocean."[70] As a result of its study, one of the principal recommendations was that the disposal of all packaged wastes be in regions where water depths were at least 1000 fathoms. This would insure that it would not be dumped on the continental shelf where the bottom is normally fairly hard but, instead, would go into the greater depths where the containers would sink into the soft sediment. At the time of writing there was no authority for the control of dumping of radioactive wastes on the high seas.

Specifications were given for the packaging of material. For example, the packages must be such that they could not be damaged or broken and would reach the bottom without rupture or appreciable loss of contents. Also the container with the material should be free of voids and have a minimum density of the order of 10 lb per gallon. It was also necessary that it have sufficient shielding for safe storage, shipment, and handling and be of such a size as to be handled quickly and conveniently. Records had to be kept as to the nature and identification of all packages that were dumped.

Provision was also made for bulk disposal operations in which the wastes were dumped directly into the sea in unpackaged form. Under these conditions it was considered preferable that the material be in such a form that it would sink when discharged, for example, in the form of insoluble beads, pellets, or briquettes. Under special circumstances the waste might also be in the form of liquid materials having a density greater that 1.1 g per cubic centimeter to facilitate mixing.

The disposal of radioactive wastes through offshore pipelines was considered undesirable.

The report included an outline of the oceanographic characteristics which might influence the burial techniques. This included items such as the chemical composition of the sea water, its physical properties, currents and mixing, and the general biology of oceans. Attention was drawn especially to the pollution aspects of radioactivity dumping and pointed out some of the problems and responsibilities that would result. It emphasized that the introduction of radioactive wastes into the ocean was an entirely new practice and unless it was conducted properly could result in new kinds of pollution. It was felt, therefore, that the agencies responsible for radioactive waste disposal should see that such practices never constituted a pollution hazard.

The report dealt with general problems of the fate of radioactive materials once they had been introduced into the ocean and speculated upon their direct and indirect hazard. Numerical evaluations were given for dilutions and the consequences of different dilutions as influencing the choice of disposal method. Concern was expressed for the indirect hazards which might arise from the accumulation of potentially harmful amounts of radioactive isotopes in marine organisms ultimately used for human food.

Practices carried out in general accord with the recommendations of this committee have been watched very carefully and so far there is no evidence that any identifiable hazards have been introduced due to the ultimate breakdown of the radioactivity containers and the movement of radioactive materials into the food chain.

NCRP REPORT NO. 10

It was evident from the outset of the studies discussed above that any successful radiation control program would depend on the instrumentation of a wide variety in sophistication to meet the need of the other committees as well as to provide general guidance. A Committee on "Radiological Monitoring Methods and Instruments" was set up under Dr. Howard L. Andrews. His report was completed in September 1951.[64] Instrumentation had to be provided for all kinds of radiation and for levels of radiation covering perhaps as much as six or eight decades. In general, instrumentation could be divided into two broad categories: (1) Radiation Detectors which generally were of fairly high sensitivity and (2) Radiation Measuring Instruments which generally had to have a wide range of dose rate. Each had to be matched to the kind of radiation it would be required to detect or measure, and frequently this lead to contradictory requirements. Accordingly, it was first necessary to indicate the reasons for the different ionizing capabilities of the different radiations and then develop an understanding of the radiation characteristics of x-rays, gamma rays, beta rays, alpha rays, and neutrons, insofar as they would produce ionization, which at that time was the principle on which most of the instruments operated. General specifications for each class of instrument were set forth.

Finally, a series of recommendations for radiation survey methods was outlined for different applications. These included the monitoring of radioactive isotope operations in the laboratory or clinic, the survey of medical installations including dental, fluoroscopic, radiographic, and therapeutic. Included was measuring equipment for industrial installations, television receivers, electronic tubes, x-ray diffraction units and electron microscopes and industrial x-ray and radium inspection systems. General recommendations were made outlining the general principles for surveys as indicated above and the instrument requirements for them.

EXPOSURE OF THE PUBLIC

It has been pointed out that in the development of various phases of radiation protection, not very much was known about any speicfic biological effects of radiation at low doses. An exception to this began to develop with the genetics experiments of H. Muller. It was in about 1933 that Muller first reported his findings on the genetic effects of radiation on drosophila. During 1945-46 Muller's concern about the effect of x-rays on genetic material resulted in his undertaking a series of lectures under the auspices of the Society of the Sigma Xi to inform the public about the harmful effects of radiation. While it was probably the possible effects of nuclear radiations that started his concern in this area, it was, in fact, the effects of medical exposures to radiation which drew his particular attention. This is perhaps the first instance where the public had been informed on a broad scale about the possible hazards of small amounts of radiation. Another potential form of public exposure to radiation began to develop in the late 40's, at which time the Atomic Energy activities were transferred from the Manhattan District under the Army to the newly-formed Atomic Energy Commission. During the Manhattan District days, laboratories had been built in several areas: Chicago, Oak Ridge, Los Alamos, Hanford, etc. The operations of these were turned over to the newly formed Commission. Because of the staffs and facilities that they had on hand, it was appropriate that they undertake extensive studies in the biomedical field in a continuation of those which had been started earlier. This type of program was funded generously in terms of any such research programs known before and so there was the prospect of a constant output of new information on the biological effects of radiation. This output has indeed continued and while of great importance has not yet provided complete answers to many of the problems that were recognized many years before.

Among the several units into which the Atomic Energy Commission was divided was the Division of Biology and Medicine under the direction of Dr. Shields Warren. During the early life of this division attention was paid to the development of radiation protection programs and the training of personnel to handle them.

As a part of their military work, the Atomic Energy Commission was planning and carrying out a series of atomic bomb tests in the Pacific. From these tests, and the actual dropping of atomic bombs in Japan, it was evident that radioactive strontium produced by the shot was injected into the upper atomosphere and disseminated worldwide. Because this material had a long half-life (of the order of 25 years), concern was felt as to its possible effects as it gradually came down to earth. To find out about this, a program known as Project Gabriel was started to study the possible amounts and effects of worldwide fallout of strontium-90. While the report was never released, its general results indicated that at the level of testing then contemplated (1948) the hazard did not seem too great. Of course the picture in this respect was totally changed with the development of the thermonuclear weapon a few years later.

Another project of interest started by the Biophysics Branch of the Division of Biology and Medicine during that period was the development of health physics training programs under AEC sponsorship. The programs, as initially conceived, would enable a modest number of college graduates in technical fields to be selected for a year of post-graduate training at Oak Ridge and Rochester. Between the two groups, it was possible to give excellent basic training to about 60 young persons per year. These have formed the backbone of the Health Physics profession as it began to really develop at that point.

Another action of concern that greatly influenced the protection picture was the strengthening of the Atomic Energy Commission under the Atomic Energy Act. While the problems of health and safety were specified in the original act, the more complete Atomic Energy Act of 1954 devoted much more attention to the Commission's responsibility for health and safety than had the initial act.

ATOMIC BOMB CASUALTY COMMISSION

Immediately after the war a farsighted program was initiated by the Atomic Energy Commission through the National Academy of Sciences to make a long-range study of the effects of atomic radiations on the populations that had been exposed during the Japanese bombings. This action led to the establishment of the Atomic Bomb Casualty Commission, an organization combining the efforts of American and Japanese scientists into a continuing study of the atomic bomb survivors.

Individual cases of nearly 100,000 survivors who had been exposed to radiation, or were in the environs of the weapons when they were discharged, have been under continuous study and evaluation, and it is probable that from this program will come some of the most useful information that we have on the effects of relatively small radiation exposures of individuals. By "small" we mean the exposures of the order of perhaps 50 to 100 rads, but this is an important region because, as events have proven, it has been possible to show certain relationships between exposure and effects in that dose range, especially that of leukemia.

Already mentioned, as starting in the period of the late 40's, were the Tri-Partite Conferences on radiation protection held between representatives of the United States, Canada and England. In all, three meetings were held: (1) Chalk River, October 1949, (2) Harwell, July 1950 and (3) Harriman, New York, in March 1953. The proceedings of these conferences were held classified at the time but they were participated in by members of protection groups from the three countries involved and, as it turns out, much of the information developed during these conferences has been fed back into the public domain. It has already been noted that the draft reports of NCRP, Committee I on External Dose and Committee II on Internal Dose, were made available to these conferences and the chairman of these committees participated in them.

At the meeting in Harwell in 1950, the Tri-Partite Conference agreed on permissible concentrations for the following radionuclides: 226Ra, 239Pu, ^{89}Sr + ^{90}Sr, ^{90}y, ^{210}Po, ^{3}H, 14C, ^{28}Na, ^{32}P, ^{60}Co, and ^{131}I. These were included as an appendix to the ICRP report of 1950 since the Harwell Meeting was attended by a number of members of the ICRP.[103] However, as noted in the report, this was not an official action by the International Commission because the Harwell Meeting followed the Commission's meetings in London. This informal action nevertheless served a very useful function in putting before the public some much needed information without waiting for formal action.

RADIATION QUANTITIES AND UNITS

Another action of broad interest in the late 1940's was the renewed interest in radiation quantities and units, whether in the area of protection or in the area of clinical radiology. It was becoming increasingly important to have quantities and units covering all the radiations then confronting us. In the mid 1930's, L. H. Gray had established the principle of energy deposition from ionizing radiation in tissue as probably the critical element in understanding the interaction of radiation and matter. In 1949, two independent groups proposed the introduction of a radiation quantity based on the energy imparted to material. The first of these was put out by Fano and Taylor in 1950.[36] The other was put out at about the same time by a British Units Committee; both were presented to the ICRU at its first meeting in London in the summer of 1950. At this meeting, the International Commission on Radiation Units (ICRU) recognized the validity of the arguments, but since there had not been time for mature consideration, postponed a decision on any specific recommendations until its next formal meeting which was to be held in 1953.

At the 1953 meeting of the ICRU, the concept of a unit based on the energy imparted to matter was adopted.[37]

An "absorbed dose" of any ionizing radiation in a material of interest was defined as the amount of energy imparted to that matter by ionizing particles per unit mass of irradiated material. It was expressed in "rads," where the "rad" is the unit of absorbed dose and is 100 ergs per gram. This step was designed to simplify the structure, both of measurement and understanding, for radiations of different kinds as, for example, neutrons, alpha rays, beta rays, and gamma rays. Since all of these could now be described or defined in terms of specific energy absorption and expressed in terms of rads, an important step forward had been made towards the quantitation of combined radiation effects from different radiations. It is true, as brought out later, that there had to be corrections for a relative biological effectiveness between different radiations but the start on the energy system of quantities and units was a substantial advance.

RECONSTITUTION OF THE INTERNATIONAL COMMITTEES

It has already been noted that at the outbreak of war at the end of the 1930's, the work of the two international committees on protection and units had gone into a state of inactivity with L. S. Taylor acting as Secretary for both. By about 1948 or 1949, plans began to generate for the resumption of the International Congresses of Radiology. It had been planned for the Fifth Congress to be held in Germany in 1940 under the Presidency of Professor H. Holthusen. It was obvious that this arrangement could not continue in 1950 and plans were made for the Fifth Congress to be held in London in July 1950 under the Presidency of Dr. Ralston Paterson.

In the meantime, Dr. Christie, who had been President of the 1937 Congress, was charged with providing for the necessary carryover from earlier

Congresses. In this connection he asked W. V. Mayneord and L. S. Taylor to jointly arrange for the reestablishment of the two international committees. Since it was realized that the great majority of the members of both committees were no longer alive, it was decided to make a clean start as far as the organization was concerned. There had been certain disadvantages as to the operating mechanism of each. For example, the Units Committee was entitled to have two members from each country represented at the Congress. This had actually meant that at one time more than 50 people attended a committee meeting. It was for this reason that in the mid 1930's the operations of the Units Committee were put in the hands of a small Executive Committee of nine people and a secretary so that, essentially, it was the Executive Committee that was responsible for the work.

The Protection Committee, having observed the difficulties that the Units Committee had had with almost unlimited membership, had from the start kept itself small. However, it was now obvious, in view of all the new radiation protection problems that would be facing the world, that the original number of five or six was too small. It was therefore agreed between Mayneord and Taylor, and subsequently approved by the Committee and finally by the Congress, that

1) The Committees would be renamed the International Commission on Radiological Protection and the International Commission on Radiological Units;

2) Each Commission would consist of 12 members and a chairman,

3) The members would be chosen primarily for their competence in the field, would be elected by the Commission itself and approved by the Congress, and

4) The Honorary Chairmanships which had been a problem earlier were eliminated. A set of specific rules was drawn up for each of the Commissions.

In assessing the work and plans of the ICRP, it is important to realize the influence of the world situation at that time. The war had been over for only five years. For more than ten years there had been little or no significant communication between the radiologists and the radiological scientists in different countries. The principal advances in radiation science and technology had been made in one country (the United States), although with heavy collaboration from England and Canada. It was not unnatural, therefore, that the dominant influence on the work of the ICRP during this early period would almost certainly be from the United States and without there being any specific intentions along this line, this is pretty much the way it turned out. Also, it should be borne in mind that in the United States it had been possible for the NCRP, with some support from the Atomic Energy Commission and others, to make a three- or four-year head start on tackling the radiation protection problems as compared with the ICRP. In fact, by the time of the reorganization meeting of the ICRP, several reports of the U. S. Committee had been published and drafts of several others were made available to the Commission for purposes of discussion.

Calling attention to this temporary position of U. S. dominance is not for the sake of credit; it was simply one of the facts of life and the situation continued with diminishing importance until nearly the end of the decade. One could probably say that beginning about 1956 there was no one national dominant influence on the ICRP with the possible exception of the Committee on Internal Emitters, which will be returned to later.

In regrouping for its production activities, the ICRP followed fairly closely the structure and procedures which had been in use for several years by the NCRP; with the main commission as the final arbiter, five subcommittees were established. The chairman of each was a member of the main commission but the members were chosen with the combined requirement of having technical competence and, to the extent possible, geographic distribution.[103]

Subcommittee I on "Permissible Dose for External Radiation" had ten members including the Chairman, Dr. G. Failla, who was also the Chairman of the corresponding NCRP Committee. In addition, there were two other members who belonged to the NCRP Committee.

Subcommittee II on "Permissible Dose for Internal Radiation" had nine members including Dr. K. Z. Morgan, the Chairman, who was also Chairman of the corresponding NCRP Committee. As for Committee I, other members of Committee II included two from the U. S. Committee.

Subcommittee III on "Protection Against X-rays Generated at Potentials up to 3-million

Volts" was under the Chairmanship of Dr. R. G. Jeager from Germany. This committee had 11 members with 2 of those from the United States, also being members of the corresponding NCRP Committee.

Subcommittee IV dealt with "Protection Against X-rays above 3-Million Volts, Beta rays, Gamma rays and Heavy Particles Including Neutrons and Protons." This was a ten-man committee under the Chairmanship of Dr. W. V. Mayneord and included two members of the corresponding committee in the United States.

Committee V on the "Handling and Disposal of Radioactive Isotopes" consisted of eight members and the Chairman, Dr. Cipriani from Canada, and this committee had one member on the corresponding U. S. Committee.

It should not be surprising, therefore, that between the two groups (NCRP and ICRP) at this point, that there should be a fair amount of overlapping of membership and that their recommendations should be essentially the same.

The next meeting of the ICRP took place in Stockholm in 1952. The occasion was a series of joint meetings between a special committee of UNESCO, the "Joint Committee on Radiobiology," the ICRP and the ICRU.[38] This was carried out under the auspices of the Radio Physics Laboratory of the Karolinska Institute and the Swedish Academy of Sciences, in both of which Dr. Rolf Sievert held an important role. The major subject under consideration was the possible genetic effects of radiation, and the meeting was attended by geneticists from several coutries. At this time it was recognized that the genetic effects of radiation were becoming better understood and should be taken into consideration in the development of any radiation protection philosophy. Figures were discussed for an average per capita dose for a large population. The proposals for an average per capita gonadal dose ranged from about 3 R as proposed by the British to about 20 R as proposed by Muller from the U. S. Actually, some even higher levels were seriously discussed but 10 R was the final value accepted informally - this, of course, being in addition to natural background radiation. However, because there was relatively little agreement between the geneticists themselves, and because of the meager information on this question, it was decided not to make any specific recommendations. (It is interesting to note, however, that subsequent recommendations made on this question in April 1956 by the ICRP, by the National Academy of Sciences in the U. S., and by the Medical Research Council in the United Kingdom gave approximately the same range of figures as those discussed at the Stockholm meeting in 1952).

Informal meetings of both the ICRP and the ICRU were held during the course of the week of the Joint UNESCO Meeting. These were largely for purposes of consolidating the organizational and committee plans which had had to be made under rather hurried conditions in London two years previously. At this time, W. Binks of U. K. was made permanent secretary in place of L. S. Taylor who had been acting secretary since 1937.

The Sixth Meeting of the ICRP was held in Copenhagen in 1953 jointly with the ICRU. For the first time these Commissions met during the week preceding the opening of the International Congress of Radiology, thus giving them more uninterrupted time to consider the problems before them. Both Commissions had most of their subcommittee members in attendance.

The report of the Commission itself, following the 1953 meeting, was brief and covered only a bare outline of the considerations. The final report was completed in December 1954 and, in addition, included the reports of the five subcommittees.[104]

The general report recommended continuation of the basic value of the permissible dose for whole body irradiation of radiation workers at the level of 0.3 R per week. In this connection they also introduced the new unit of dose, the rad, which was approved and recommended by the ICRU at the same joint meetings. Some of the dosage values were also expressed in terms of the "rem," pointing out, though, that this unit was not officially recognized by the ICRU. (It was some years later that this problem was further sorted out by the ICRU and finally recognized jointly with the ICRP.)

The Commission had found in 1950 that it was not in a position to make firm recommendations regarding the maximum permissible amounts of radioactive isotopes that might be taken into or retained by the body. Recognizing that much of the uncertainty still remained regarding the behavior of radioactive materials in the body, the Commission in 1953 recommended, with some more confidence, certain values for permissible body burdens and permissible concentrations in air and water.

The ICRP, recognizing the problems of exposure of the population to radiation, made the first recommendations for levels where large numbers of people might be exposed. These were specified as values one tenth of those accepted for occupational exposure, but it was not specified whether for groups or individuals.[104] The report dealt at some length with the question of health surveillance and personnel tests including more specific recommendations regarding pre-employment and routine blood counts and blood count techniques.

The report of Subcommittee I on "Permissible Dose from External Radiation" followed fairly closely, but in considerably less detail, the recommendations developed by the corresponding committee of the NCRP. Permissible dose was defined as "a dose of ionizing radiation that in the light of present knowledge is not expected to cause appreciable bodily injury to a person at any time during his lifetime."

As used here "appreciable bodily injury" means any bodily injury or effect that a person would regard as being objectionable and/or competent medical authorities would regard as being deleterious to the health and well being of the individual. Dose is used here more particularly as tissue dose in the radiated tissue organ or region of interest. What constitutes the region of interest depends on the conditions of exposure and must be taken into account in assigning numerical values to the permissible dose or doses applicable to a given set of conditions. This may be compared with the NCRP definition, which was essentially the same.

The report went on to define and discuss the critical organs and effects of concern. These were listed as follows: skin, blood forming organs, gonads with respect to impaired fertility, and eyes with respect to cataracts.

It was indicated that when the whole body is irridiated, a rough approximation of the dose in the organ might be made on the basis of an average or effective depth of an organ below the surface of the skin. For purposes of uniformity in calculation the following effective depths were suggested: skin (basal layer of the epidermis) - 7 mg/cm^2; blood forming organs - 5 cm; ovaries - 7 cm; testis - variable, depending on conditions of exposure, minimum 1 cm; lenses of eye - 3 mm.

For whole body exposure and exposure of the critical organs, a limit of 600 mR per week was recommended for the basal layer of the epidermis over a significant area. For all others, the values were recommended at 300 mR per week averaged over a significant area or volume.

For purposes of the report, the rem was defined as the absorbed dose of any ionizing radiation which has the same biological effectiveness as one rad of x-radiation with an average specific ionization of 100 ion pairs per micron of water. In terms of its air equivalent in the same region, the dose in rems is equal to the dose in rads multiplied by the appropriate RBE. Following this definition, a table of values of RBE was given.

The report of the Subcommittee II on "Permissible Dose for Internal Radiation" was essentially the same as developed by the corresponding committee of the NCRP.

As already noted above, the program on internal emitters had been started in the U. S. by the NCRP and through the urging of the NCRP to the AEC was proceeding mainly at the Oak Ridge National Laboratory with increasing support from the AEC. It was not uncommon, when holding meetings of the NCRP Committee II, to invite members from the corresponding ICRP Committee to meet with them and vice versa. As a consequence, it has never been possible to make any real distinction between the recommendations of the two groups. Therefore, for discussion of the ICRP report in this area reference should be made to the NCRP discussion already given above.

When Committee III on "Protection against X-rays Generated at Potentials up to 2 MeV" undertook its task, it started with the report of the corresponding NCRP Committee as a base. Except for the fact that for international applications it was desirable to express some of the recommendations in more general terms, the ICRP Report again followed very closely that of the NCRP, which also was discussed in detail above.

By the time of the close of the meetings in Copenhagen in 1953, Committees IV and V had made little progress towards the development of a final set of recommendations. Their one-page combined report endorsed the permissible level of 300 mrad per week for quantum energies above 3 MeV. In addition, they made the tentative recommendation that the permissible exposure to neutrons in the energy range of 0.025 eV (thermal) to 1 MeV shall be either (1) that exposure which shall produce an energy absorption of 30 mrad per week at a depth of 2 cm below the surface of the tissue or, (2) exposure for periods not exceeding 40 hours per week to specified neutron fluxdensities. This was followed by a table giving, essentially, permissible neutron fluxdensities as a function of neutron energy. They also gave a table of tentative RBE values: 10 for fast neutrons and protons up to 10 MeV, 10 for naturally occuring alpha particles, and 20 for heavy recoil nuclei.[104]

NCRP ACTIVITIES 1953 TO 1956

The period 1953 to 1956 was one of considerable activity for both the NCRP and the ICRP. This section will deal briefly with five reports put out by the NCRP during that period. Another subcommittee split off from the original Subcommittee V was a new one to deal with the problem of "Protection Against Radiations from Radium, Cobalt-60 and Cesium-137" (Report No. 13). This was under the Chairmanship of C. B. Braestrup, and its first report was issued in September 1954.[67] In part, this report, in its recommendations, was an updating of the radium protection reports issued in 1934 and 1938 (Reports No. 2 and 4). However, since that time, encapsulated sources of cobalt-60 and cesium-137 had become available and because their use for therapeutic purposes was very much the same, they were treated together comprehensively in one report. While no specific references were made to industrial applications the basic principles and attenuation data that were developed by this committee were applicable to both medical and the industrial uses.

Since the fundamental philosophy and criteria for protection had been developed earlier by Committee I, this report, like the one for x-ray protection, was more in the nature of a code of practice. It gave, in considerable detail, recommendations on the equipment and facilities for the handling, storage, and transportation of sealed sources of radioactivity in the intermediate range— that is between the microcurie and kilocurie sizes. The sections on medical applications: interstitial, intracavity and surface; and medical applications: teletherapy, dealt with the precautions necessary both during the operative stages and the stages while the source might be in or on the patient. Special emphasis was put upon the careful rehearsing of all operations to insure that the procedures were carried out in the most expeditious manner possible.

By the early 1950's the use of large sources of radioactivity for external beam use on patients was being introduced. This technique, generally described as teletherapy, might involve quantities of material in the kilocurie range. Actually, the art was not far along at this stage so that the recommendations were rather rudimentary, but it was possible to adequately specify and measure the leakage radiation permitted.

A few non-medical applications were considered. These included sources used for either calibration or checking the calibrations of clinical survey instruments. Also included were sections on research applications and the use of special sources in Civil Defense activities where very rough usage could be expected.

Considerable attention was devoted to the problems of protection surveys, personnel monitoring, and working conditions. These followed fairly closely the pattern developed for high energy x-ray uses.

Conditions that might result either from accidental spillage or fire might become of serious proportions considering the amounts of radioactive material involved, and as a consequence this subject was dealt with. Experience has shown that most of the broad recommendations in the report are still valid and generally used except that there have been some changes in the basic permissible dose levels for which corrections need to be made.

The appendices were rather more complete than for most reports including a summary of all the various shipping rules, barrier design data and computations, and many curves and tables giving barrier requirements as a function of radiation quantity, distances, etc.

Since 1940, a whole new class of radiation sources had resulted from the rapid development of such high energy electron accelerators as linear accelerators, electron cyclotrons, electron synchrotrons, and betatrons. Of this group the betatrons and electron synchrotrons were being built commercially in the late 1940's. Because these accelerators are sources of high energy radiations and because of their widespread applications, they represented a potential hazard to operating personnel, to patients in hospitals, and to the public. Therefore, attention to this particular class of devices was undertaken as one part of the task of Committee V under the Chairmanship of Dr. H. W.

Koch. (As already noted most of the other problems of original concern to this committee had earlier been studied by other committees.)

The first report of this committee, Protection Against Betatron - Synchrotron Radiations up to 100-Million Electron Volts, was issued in February 1954[68] (Report No. 14).

The report attempted to supply the necessary recommendations for protection of the operators and outsiders. Only betatron and synchrotron sources were considered since adequate data were lacking for most others. The hazards resulting from the various radiations produced by these sources were included as well as those due to associated effects such as noise, electricity, and ozone production.

It was recognized that the experimental data related to protection requirements for betatrons and synchrotrons were far from complete. Measurement techniques and systems of units were not well established because there was a general deficiency of a standard measurement procedure in the high energy field; consequently, complete appendices were given to provide detailed discussions of proposed units and measurement procedures. These details are not ordinarily required in a protection program but, as indicated, were necessary here because of the novelty of the whole field.

The following issues received special attention:

1. The choice of the units for radiation intensity (Watts/cm^2) and for dose (ergs/gram).
2. The choice of a secondary standard of irradiation intensity.
3. The choice of a 5-cm thick lucite cover for the sensitive element of a survey instrument.
4. The recommendation, for high energy installations, of the standard personnel monitoring procedures then in use below 2 MeV.

While it had been some years since high voltage electrical protection was a matter of concern in the general protection field, it had become one of considerable concern in most of the high energy accelerators. This was in part because of their general experimental nature and the fact that many of their features were far from standardized.

Because much of the protection cost depends on the plans for the overall installation, attention was directed to structural details and layout plans which would minimize costs and maximize shielding. Since the requirements might be different between medical and industrial installations, recommendations were given for both. The general procedures for radiation protection surveys and monitoring and for working conditions are very much the same as for other radiation installations, with the exception that the measurement procedures might have to be more elaborate in order to insure that any error in radiation measurement would not end the dangerous direction, i.e., such as to increase the risk to people, if in error.

The next report completed by the NCRP dealt with the problems of Safe Handling of Cadavers Containing Radioactive Isotopes. This was managed under the Chairmanship of Dr. E. H. Quimby and the report was issued in October 1953[69] (Report No. 15).

With the rapidly expanding uses of radioactive material for diagnostic purposes, either in sealed sources or by ingestion, it became obvious that awkward situations might arise shortly after a patient might have received a large internal therapeutic dose of radioactive material. The handling of a body might pose serious problems of radiation exposure for the pathologists and for the embalmers, especially under conditions where there has been little experience with such a body. It was felt important for members of these groups to realize the existence of the problem and know how to meet it and, equally important, that they not exaggerate the hazard nor be unreasonably fearful when it is minimal or nonexistent.

It was recognized at the time that during the next few years the probability that such bodies would be encountered frequently was very low. As a matter of fact, while the frequency has of course increased, it has not done so to the point of posing special difficulties simply by virtue of the number of cases. Attention had to be given to the choice of individuals who might work with the bodies. For example, in the case of death at a particualr hospital, it might frequently be the same pathologist who had to perform autopsies on all of the bodies. On the other hand, it was unlikely that the same embalmer would be likely to have to prepare all of the bodies for burial.

The report pointed out that the maximum permissible exposure for any handlers would never be reached in dealing with a body of an individual who had received only a tracer dose of any radionuclide, such as likely to be used for diagnostic purposes. It was thus primarily the therapeutic doses that might be of concern. Because of the fact that all patients having a body content of more than about 30 millicuries of radioactivity would be required to stay in the hospital, it was not likely that cases that die outside the hospital would be encountered very often by funeral directors. If such a body were embalmed without opening, using the standard aspiration and injection method, the hazard would be minimal.

On the other hand, if a patient containing the radioactive material died in the hospital it is highly probable that an autopsy would be performed, in which case the radiation hazard could be significant. Offsetting this would be the fact that the institution would normally be equipped for handling large amounts of the radionuclide in question. It would also have an official radiation safety officer and other facilities so that while the problem of autopsy, embalming, etc. could be complicated, it was nevertheless manageable.

The report also dealt with some of the extreme, but rare, cases that might arise where the patient would die shortly after injections of large amounts of gold-198 or iodine-131 under which circumstances an immediate autopsy would be inadvisable. The report outlined varying combinations of circumstances and how to deal with them. Finally, it gave the recommended form for a radioactivity report that should accompany any body containing radioactive materials submitted for autopsy or embalming.

By the mid 1950's the NCRP had reached a point of seeing the necessity for revising or updating some of its reports that even then were less than ten years old. The first of these was the development of a new report on x-ray protection to replace Report Number 6. The new report, issued in December 1955, entitled X-ray Protection (Report Number 18)[72], was prepared by a committee chaired by Dr. H. O. Wyckoff but with some changes made from the original membership.

The contents of this report will not be outlined since it was primarily an updating of the earlier one. However, it is worth mentioning that the report did give increasing attention to the exercise of caution and restraint in the use of x rays on patients and urged that current methods and practices be reviewed to see if the same results could be obtained with less radiation. Emphasis was placed on minimization of the use of student technicians as radiographic subjects for training purposes, and that the student's body or any part should not be exposed to more than the appropriate maximum permissible weekly doses. Also, continued emphasis was given to urging the minimization of radiation uses for the treatment of benign conditions, especially when the patients might have been occupationally exposed.

NCRP REPORT NO. 19

It had been obvious for some years that one method of controlling radiation exposure might be through the regulatory or licensing procedure; the subject had been brought up informally from time to time within the NCRP discussions. While the NCRP had never taken any official position one way or the other with regard to licensing and regulation, it nevertheless recognized that this type of control was almost certain to come. It was therefore considered as a fact of life, and in early 1953 a Committee under the Chairmanship of L. S. Taylor was organized to deal with the problem. Its report was completed in July and issued in December 1955.[73]

The concept of radiation regulation was not especially new and was probably further advanced in several European areas than in the United States, since Europeans traditionally have tended more toward the regulatory technique. Perhaps the earliest such actions in this country were contained in the amendments of the Chicago Code of 1911 through the ordinances of April 7, 1920, and April 8, 1921. Also, in New York there were amendments to the New York Regulation (concerning x-ray laboratories) July 24, 1923, and February 26, 1926. One of the more important and comprehensive plans for radiation regulation was developed in Sweden and put into action in July 1941.[39] There were probably other examples, but no attempt has been made to seek them out; it just happens that these had come to the attention of the writer.

The report prepared by the NCRP committee represented something of a departure from the recommendations on radiation protection that it had prepared in the past. For the most part, each of the NCRP reports prior to this one had presented technical facts and data, and from these developed certain specific recommendations as to methods and procedures for the achievement of adequate protection of persons against the harmful effects of radiation. This new study presented the problem of radiation in relation to its possible control by state or municipal authorities. The problem was recognized as a new one to all except possibly two or three states and, at the time the study was started, only one state had a set of comprehensive regulations designed to control all forms of ionizing radiation. A few others were in the process of developing regulations; at the present time many states have detailed regulations.

While, as noted, the NCRP had no policy position one way or the other with regard to the use of regulations as a means of controlling radiation hazard, it was probably the accepted but unwritten policy, that it discourage the incorporation of *its* recommendations directly into legislation or other similar control acts. It was felt that better results could perhaps be obtained through education and voluntary compliance. However, by the early 1950's, it recognized that conditions were changing rapidly and that state control might become a necessity with the accelerating growth of radiation uses.

It was early in the 1950's that the NCRP adopted the position that it would not recommend or oppose the incorporation of its findings into state codes. However, if a state decided of its own volition to develop radiation control legislation or regulations, the committee agreed that it would provide any assistance within its capabilities to the end that the regulations be made as sound and workable as possible. A further objective was to assist the states in the matter of assuring the maximum degree of technical and operational uniformity between their several radiation regulations.

The NCRP studies of radiation control problems, resulting in its report on Regulation of Radiation Exposure by Legislative Means, were begun early in 1953 as a result of a joint request by representatives of the American College of Radiology, The American Medical Association, and the United States Public Health Service. This request was in turn prompted by repeated inquiries by individuals and groups seeking guidance and information on the subject.

Another procedural departure in the preparation of this report was the fact that before final publication its drafts were circulated for comment to many states and to more than 100 attorneys, engineers, radiologists, and general scientists outside of the strictly radiation field, from whom many helpful comments and suggestions were received.

At the same time that the NCRP was working on its study, the Atomic Energy Commission was developing regulatory procedures which they expected to have to bring into play. There was close coordination between the NCRP committee and the AEC in their development of radiation regulations under the Atomic Energy Act of 1954. The principal difference between the Federal regulations and those of the NCRP lay in the fact that the AEC regulations pertained only to radiation sources as defined by the Atomic Energy Act, whereas the NCRP regulations covered all sources of radiation including radium, x-ray machines, and high energy accelerators. No attempt was made in the report to resolve the jurisdictional problems that would almost surely develop between federal, state, and municipal authorities in their efforts to control or regulate the use of ionizing radiation.

In addition, it was necessary to write the report in a language as non-technical as possible since it would undoubtedly be used by many people not familiar with the language, let alone the concepts. After stating the problem in simple terms, three possible solutions were pointed out. The least restrictive one would be to follow the then current broad policies augmented by a widespread program of education. A second would be one in which all sources of radiation are registered. The third and most restrictive plan would be one calling for inspection and licensing of all radiation sources. As things stood, the second solution seemed to be the one best suited for general use although, by law, much of the material coming under the Atomic Energy Act would be subject to both licensing and inspection.

After this introduction there was a pro and con discussion of the philosophy of radiation protection legislation regarding its needs, questions of enforceability of radiation legislation, licensing versus registration, etc. There were also problems of how to constitute a radiation regulatory body with the development of the necessary advisory services and radiation protection experts. Finally, there was a listing of the critical items that should be included in existing or proposed radiation control acts. These would be: (a) policy, (b) definitions, (c) scope, (d) exemptions, (e) prohibitions, (f) registration, (g) advisory boards, and (h) administration.

Two appendices were included and comprised, in fact, the bulk of the report. The first of these gave a suggested State Radiation Protection Act meeting the critical items outlined above and included sections on hearings, penalties, injunctions, conflicting laws, existing rights and remedies, severability, etc.

The second appendix gave a draft of some suggested regulations which it was hoped might serve as a guide and lead towards greater uniformity between the different bodies that might be developing regulations. The proposed regulations were divided into 15 sections as follows: (1) scope, (2) applications, (3) definitions, (4) exemptions, (5) standards, (6) registration, (7) maximum permissible dose, (8) personnel monitoring and area radiation surveys, (9) radiation exposure records and reports, (10) responsibility, (11) storage of radioactive materials, (12) radioactive contamination control, (13) radiation information labeling, (14) disposal of radioactive wastes, and (15) technical standard guides and general information to be used in achieving the requirements of the regulations.

While this report was in its final stages of drafting, many states and municipalities learned of its existence and sought copies. There was extensive interest in its content and within a relatively short time the general pattern was being discussed and introduced at many governmental levels. The Federal Regulations developed by the Atomic Energy Commission were quite similar in principle. Shortly after the issuance of the report, the Council of State Governments began to concern itself with state regulations of radiation exposure and they ultimately developed their own pattern for proposed legislation and regulation. They were, generally speaking, much more experienced in the legal and administrative aspects of the problems than the NCRP could have been expected to be, so there were some changes introduced in the process. However, the general pattern and the basic requirements remained essentially as developed by the NCRP committee.

A few years later, the NCRP felt it should review the status of the report and the status of the regulatory field and make a decision whether to continue with that kind of activity or leave it to the Council of State Governments, the individual States, and the Atomic Energy Commission. It was ultimately decided to go no further with development of such recommendations unless, at some time in the future, it became evident that unacceptable conditions were arising because of inadequate guidance of the type that might be provided by the Committee. That time has not arrived and now does not seem likely to.

OTHER DEVELOPMENTS 1950 to 1955

The most important single radiation event in the 1950's was almost certainly the detonation of the first thermonuclear device at the test site in the Pacific in which the natives of Rongelap and a few Japanese fishermen received radiation exposure. The generation of a large quantity of radioactive debris had been anticipated, although perhaps not to the extent in which it was actually produced. Because of this, what were considered to be reasonable preparations were made for evacuation or control of the area in which fallout might occur. The exposure of the people noted above was obviously unexpected and drew almost immediate worldwide attention to the potentialities of this new type of weapon. While the general public had certainly been passively aware of the weapons testing programs in the Pacific and the potential hazards from the radiation resulting therefrom, the results from the thermonuclear tests were startling and alarming. For the first time there was an acute public awareness and public concern about the radiation hazards, whether from normal testing or from the much worse situation where such weapons might have been used against man in anger.

The problem was further exacerbated by a rash of spectacular press and magazine writings and books, few or any of which were reasonably in accordance with the facts and virtually all of which were sensationalized in the extreme (I believe it is worth noting that some of this was undoubtedly brought about because of the government's unwillingness, for military security or other reasons, to discuss and expose freely what was actually going on. Be this as it may, the fact still remains that the sensational and inaccurate writer can still sell his stories and have them published, whereas the cautious and generally more accurate writer can find no outlet for either his direct articles or his rebuttals to the sensationalized ones.)

Spurred by world concern, the First Atoms for Peace Conference was organized and held in Geneva in the summer of 1955. This consisted of papers of both general and specific technical nature and extensive exhibit materials and really represented the first time that large numbers of radiation scientists had gathered to discuss the problems and implications of the development of atomic and nuclear energy. Scientists from behind the so-called Iron Curtain were present and there was relatively free intermingling and exchange of ideas with them and with all others. As far as protection is concerned, the program included some dozens of papers dealing with all phases of radiation protection. Nothing startingly new was introduced but it was learned that all of the countries in the world were using the basic radiation protection standards that had been recently recommended by the ICRP.

One of the more interesting points was brought out by the Japanese statement that they had abandoned the idea of trying to establish a causative relationship between dose and effect within the permissible occupational exposure limits for purposes of workmen compensation. Instead they had arbitrarily adopted the principle that if an individual had worked with radiation and at some later time developed any disease which might reasonably be ascribed to radiation at any level of exposure, it was presumed that the disease was caused by the exposure and the person was compensated accordingly.

The announcement was made by both the USSR and the U. S. that their respective countries had produced electrical energy through the use of a nuclear reactor as a heat source. This, of course, was the opening of a whole new kind of radiation hazard and protection problem.

A year or two before the Atoms for Peace Conference, and in an effort to provide more world-wide cooperation in the nuclear field, steps had been taken to establish some kind of an international agency to act as a focal point. This had resulted, in September 1957, in the establishment of the International Atomic Energy Agency located in Vienna. It was still virtually in the process of organization during the second Atoms for Peace Conference. Of special interest to this report is the fact that they planned from the outset to put substantial emphasis on matters of radiation hazard and protection.

Dr. Detlev W. Bronk, then President of the National Academy of Sciences, organized an informal meeting with a number of Americans attending the Atoms for Peace Conference. The purpose of this meeting was to discuss the possibility of setting up a large-scale study, within the structure of the National Academy of Sciences, on the Biological Effects of Atomic Radiation (BEAR). It was indicated that this would be financed in a very substantial way by the Rockefeller Foundation. The study was organized in the fall of 1955 and issued its first report in June 1956.

By the mid 1950's there was a growing demand for more professionally trained people to assist in radiation control problems. This had already led to establishment of Health Physics Training Courses as note above. But, in addition, there was a growing feeling among these people that they should be attached to some kind of professional organization where they could submit papers, exchange viewpoints, and carry out other activities associated with their professional needs. Health Physics by its very nature includes a wide range of technical disciplines and it was obviously not easy to fit such a profession into any of the existing professional organizations which were primarily medical or primarily biological or primarily physical or chemical in their individual natures. This was recognized by K. Z. Morgan and others, and through their efforts the Health Physics Society was organized. This has been a useful, productive, and active organization and it has continued to thrive and fill a need that it did not seem possible to otherwise fill.

The first half of the 1950 decade closed with another significant event, namely, the establishment of the United Nations Scientific Committee on the Effects of Atomic Radiation (UNSCEAR). This was done by the adoption of a resolution by the General Assembly on December 3, 1955.[1,2,3] Its charge to UNSCEAR included the following: (1) to receive and assemble, in appropriate and useful form, radiological information on (a) observed levels of ionizing radiation and radioactivity in the environment and, (b) scientific observations and experiments relative to the effects of ionizing radiation upon man and his environment; (2) to recommend uniform standards with respect to procedures for radioactivity sample collection and instrumentation and radiation counting procedures; (3) to compile and assemble in suitable manner various reports on observed radiological levels; (4) review and collate national reports on scientific observations; and (5) make yearly progress reports on radiation levels and radiation effects on man.

The United Nations Committee held its first session in March 1956, at which time it elected Dr. C. E. Eddy of Australia as its Chairman.

As a result of its first session, the committee

decided to examine problems falling within its field of competence under the five general headings: (a) Genetics, (b) Effects of irradiation by internally absorbed isotopes and the effects of external radiation, (c) Natural radiation levels, (d) Exposures during medical procedures and occupational exposures and, (e) Environmental contamination.

In the establishment of the United Nations Committee it should be borne in mind that the main impetus for doing this was to consider the problem brought on by world-wide testing of nuclear weapons and, of course, the results that might be expected from the weapons already discharged. At the risk of oversimplification, their first findings seemed to surprise everyone although the basic elements of these findings had been known to a modest number of people for a number of years at least. These findings were (1) irradiation from natural sources varied over a range of some threefold under widespread conditions of habitation and as much as tenfold for a few special locations; (2) radiation exposure from fallout was usually only a small fraction of that from natural sources, when considered on a world-wide basis; (3) average irradiation of the population from medical procedures was a substantial fraction of irradiation from natural sources and in some countries possibly almost equal to it; (4) any attempt to evaluate the effects of sources of radiation to which the world population is exposed can produce only tentative estimates with wide margins of uncertainty.

1956 - A YEAR OF CHANGES

Perhaps the most significant actions influencing the philosophy and management of radiation protection were the two studies carried out by the Medical Research Council of Great Britain and by the National Academy of Sciences in the United States.[40, 171] Both of these studies, carried out on a crash basis, dealt with the biological effects of atomic radiation (as it was then called) with special emphasis on the genetic effects of radiation. Reasons for the urgency almost certainly included the prospect of thermonuclear testing to be carried out in the near future, and a response to the very considerable emphasis being lent to the subject by the first Atoms for Peace Conference, the United Nations Scientific Committee, and the establishment of the International Atomic Energy Agency. National scientific backing was necessary for all of these.

While there was an "awareness" between the Medical Research Council and the Academy of Sciences, the two groups essentially worked independently and it was important to note that while expressed in quite different ways, the basic conclusions of both groups were essentially the same. They agreed to a simultaneous release of their reports.

The prime conclusion was that the principal limitation on radiation exposure of people should be set by the genetic hazard involved. All radiation exposures were assumed to be cumulative, there was no cellular recovery of any radiation genetic damage, and there was essentially a linear relationship between dose and effect down to zero dose.

These conclusions were based primarily on the drosophila experiments by Muller who was a member of the NAS Committee. The conclusions of the British Committee were essentially the same but not expressed in such quantitative terms as was the Academy report.

One of the principal recommendations was that no individual radiation worker should receive more than a total cumulative dose to the reproductive cells of 50 R to age 30 and not more than an additional 50 R up to age 40. At those ages over half or over nine tenths of their children will have been born, respectively. For practical purposes this meant limiting the occupational exposure of individual radiation workers to 5 rems per year. It might be noted that before reaching this final recommendation the committee had inquired of the Atomic Energy Commission what range of practical radiation levels were being reached in normal operation. It turned out that because of the normally cautious procedures of the Commission it was a rare occurrence for anyone to exceed more than about one tenth of the permissible occupational exposure (approximately 15 rems per year) allowed up to that time.

As "discovered" by various other groups both before and after these particular studies, there was apparently some element of surprise in finding that the exposure for medical procedures greatly exceeded those experienced or projected from radioactive fallout.

It might be noted in passing that the same questions were reevaluated by the committees of the National Academy of Sciences in 1960,[181] by which time other experimental work in the genetics area had begun to become available. As a result of the study, their recommendations which were sharp and positive in 1956, were much more carefully conditioned in 1960. By the late 1960's, it was generally accepted that some of the basic concepts of the 1956 recommendations were wrong as far as generalizations were concerned.

ICRP MEETINGS - (APRIL 1956)

The next regular meetings of the ICRP and ICRU were scheduled to be held in 1956 in advance of the 7th International Congress of Radiology being held in Mexico City. Because the programs of both commissions were so heavy and to a considerable extent interrelated, it was decided that they would meet jointly for a two-week period in Geneva. Several important steps were taken by the Commission at this meeting. In its report, prepared in 1953,[104] the permissible dose and the permissible weekly dose were defined in precisely the same language used by the NCRP (H59, page 27).[71] It was evident that in the light of many additional discussions, and since this was originally phrased in 1949, the definition should be sharpened. This was accomplished at the Geneva Meeting:

"The permissible dose for an individual is that dose, accumulated over a long period of time or resulting from a single exposure, which, in the light of present knowledge, carries a negligible probability of severe somatic or genetic injuries; furthermore, it is such a dose that any effects that ensue more frequently are limited to those of a minor nature that would not be considered unacceptable by the exposed individual and by competent medical authorities. Any severe somatic injuries (e.g., leukemia) which might result from exposure of individuals to the permissible dose would be limited to an exceedingly small fraction of the exposed group; effects such as shortening of life span, which might be expected to occur more frequently, would be very slight and would likely be hidden by normal biological variations. The permissible doses can therefore be expected to produce effects that could be detectable only by statistical methods applied to large groups."

The Commission was aware of the forthcoming recommendation by the National Academy of Sciences that occupational exposure or exposure of any individuals be limited to the order of 5 rems per year. Realizing that it would be a foregone conclusion that this number would be adopted if it was so recommended, it was tentatively put forward at the Geneva meeting in April 1956. It received official approval of the Commission before the publication of the report in 1958.[105]

In recognition of the increasing necessity for controlling the exposure of people other than radiation workers, the Commission defined several groups or categories of exposure as follows:

A. Occupational exposure
B. The exposure of special groups, (a) Adults who work in the vicinity of controlled areas but who are not themselves employed on work causing exposure to radiation, (b) Adults who enter controlled areas occasionally in the course of their duties but are not regarded as radiation workers, (c) Members of the public living in the neighborhood of controlled areas.
C. Exposure of the population at large
D. Medical exposure

The subdivision of the exposure of special groups, B, was of particular interest to the British and was not utilized by most countries. It has since been dropped by the ICRP.

At its meeting in New York in the spring of 1958, the ICRP[107] adopted the age-proration principle which had been adopted in 1957 by the NCRP.[74] Also, the once-in-a-lifetime exposure of 25 rems for accidental high exposure and the emergency exposure of 12 rems, which had also been adopted by the NCRP in 1949, were now adopted by the ICRP. Following the New York Meetings, an ad hoc Publication Committee, under the chairmanship of Dr. Failla, did the final checking and editing of the report.

Of special importance was the recognition of the population exposure problem. The ICRP discussion of the genetic dose problem was expressed in somewhat tentative terms because it was obvious that no group had yet had time to really evaluate the British and American reports in depth. In concept, a permissible genetic dose was defined as that dose, which if it were received by each person from conception to the age of childbearing, would result in an acceptable burden to the whole population.

The report then went on to indicate that the genetic dose to the population could be assessed as the annual genetically significant dose multiplied by the mean age of childbearing which, for the purpose of their recommendations, would be taken as 30 years. The annual genetically significant dose to a population is the average of the individual gonad doses, each weighted for the expected number of children conceived subsequent to the exposure. It concluded, *"It is suggested* that the genetic dose to the whole population from all sources additional to the natural background should not exceed 5 rems plus the lowest practicable contribution from medical exposure." This meant 5 rems distributed over a 30-year period, which is about half of the level of 10 rems suggested by the NAS, but including medical exposure which had earlier been estimated to average some three or four roentgens over a 30-year period.

The Commission, then recognizing the validity of the argument that certain sources of exposure might claim priority, *suggested* the possibility of apportionment of the total and gave an illustrative example. It was stated clearly that these were just suggestions and examples, but nevertheless large numbers of readers took the example as a recommendation and for a while strongly promoted the apportionment principle as outlined.

There were numerous other important details in the 1956-58 Report since it was the first one for which there had been an extended period of time for study and adequate meeting time.

The other important action taken in 1956 was the general reorganization of its subcommittees, which had been initially organized in 1950 and changed somewhat in 1953. In the meantime, it had become obvious that their Committee I, Permissible Dose for External Radiation; Committee II, Permissible Dose for Internal Radiation; and Committee III, Protection Against X-rays up to Energies of 3 MeV and Beta Gamma Rays from Sealed Sources had all functioned effectively and issued significant reports. The remaining committees had functioned virtually not at all so their chairmen and membership were changed in 1956.

Still another action involving the ICRP occurred in 1956 when it and the ICRU received an invitation from the United Nations Scientific Committee on the Effects of Atomic Radiation to cooperate in its work involving exposure from medical procedures. The ICRP and ICRU held a special joint meeting in New York in the fall of 1956 to consider this invitation and ended by accepting the assigned task and submitting a report to UNSCEAR that was subsequently published in October 1957.[106] It was at the time of this joint meeting that the commission held further discussions of its Geneva recommendations made earlier in the year. Still further discussions were carried out at the special meeting of the Commission, this time held in New York in the spring of 1958, at which time some relatively substantial changes were introduced into the 1956 report which then had to be resubmitted to the whole commission for approval. Final review and draft, following extensive work by an Ad Hoc Publication Committee, resulted in the finished report. These activities were thus blended with the special ones being carried out by UNSCEAR.

CHANGE OF MPD BY NCRP - 1956

By the mid 1950's, the feeling was beginning to generate that the basic recommendations of Committee 1 should perhaps be re-examined even though they had only been in published form for a year or two. However, no particular action was taken until issuance of the report by the BEAR Committees of the National Academy of Sciences essentially recommended the reduction of the basic occupational exposure from a level of 15 to 5 rems per year. At this period the original Committee 1 had been largely disbanded so the Executive Committee of the NCRP decided to expand its own membership and undertake the task itself. This was done in a series of short meetings during the fall of 1956. Early agreememt was reached on the general acceptability of 5 rems per year but the members were concerned that with such a relatively low value, the chance of occasional overrun would increase and might be followed by unreasonable penalties.

Over the years past, no evidence had developed that individuals exposed at levels, which might have been as high as 15 rems per year, had suffered any discoverable injury; actually most installations were substantially below this.

Numerous efforts were made to find a mechanism for setting up some kind of a reasonable averaging mechanism so that if there was a small overrun in one year, it might somehow be averaged out the following year provided that over some limited period, like three or four years, there would be no total overrun. The question was disucssed with "regulators" (of which there was now a substantial stock) and lawyers who had had experience in state laws and regulatory matters. They all opposed any such concept.

However, out of the discussions there developed the concept of age proration whereby an individual's total exposure could be related to his age at the rate of, say, 5 rems per year. This was expressed in the now well known formula 5(N-18). This idea was proposed to the regulators and lawyers and they grasped it at once as satisfactory. With this encouragement, the question was put before the entire membership of the NCRP and in November 1966, the concept was approved.[74]

After publication of the initial recommendation developed by the Executive Committee, the NCRP decided to study the question further and under conditions of somewhat less urgency. This was done during the next year's time and the final recommendations, which differed only in small detail from the original, were adopted, published in April 1958, and added as a supplement to all existing handbooks.[77]

As already noted, the age proration principle was also subsequently adopted by the ICRP at its meeting in the spring of 1958.

At least as far as the public press was concerned, and indeed for some of the technical people, there seemed to be a tendency towards misunderstanding that the real impact of radiation exposure on the genetic system embodies exposure averaged over the population, or at least over the parts of the population where crossbreeding was expected. To emphasize this, the first new recommendations (February 1957) expressed the population dose in terms of man-rems per million people. The figure tentatively put forth was 14 million man-rems per million of population that might be accumulated up to age 30. It included 4 million man rems for medical exposure at the levels that were believed to have been required at that time. That general type of recommendation was not included in the final recommendations (January 1958) although it has never been withdrawn as a concept; it has some limited usefulness.

The 1957-58 special reports of the NCRP dealt briefly with the problem of exposure of persons outside of controlled areas. This was a question that had not been dealt with previously by the NCRP. At this time they adopted the principle that "the radiation or radioactive material outside a controlled area attributable to normal operations within the controlled area, shall be such that it is improbable that any individual will receive a dose of more than 0.5 rem in any year from external radiation."

It further specified that "the maximum permissible average body burden of radionuclides in persons outside of the controlled area and attributable to operations within the controlled area shall not exceed 1/10th of that for radiation workers." It was emphasized in both cases that the exposures, body burdens, and concentrations, as applied to persons outside of controlled areas, might be averaged over periods up to one year. The report also included special allowances for overruns provided that certain levels were not exceeded within a 13-week period. It emphasized that such doses should not be encouraged as a part of routine operations but should be regarded as an allowable but unusual situation. Finally, it was recommended that except for planning or convenience of calculation design or administration, the NCRP would henceforward discontinue the use of either the weekly MPD or MPC.

ICRU - ICRP STUDY FOR UNSCEAR

Returning to protection activities at the international level, in November 1956, the ICRP and the ICRU received joint invitations from United Nations Scientific Committee on the Effects of Atomic Radiation for cooperation in one phase of the UNSCEAR work involving the exposure of the population from medical procedures. The main objectives of the joint study were: (a) to consider and discuss the question of how to arrive at reliable data indicating the doses to different parts of the body (particularly the gonads) received by individuals, and in the aggregate by large population groups, due to the medical use of ionizing radiation, (b) to examine what recording system, if any, is at present feasible for the determination of the relevant dose values, and (c) as a result of the studies to submit to the Scientific Committee on the Effect of Atomic Radiation as soon as possible, and in any event before September 1, 1957, a report upon their deliberations and conclusions on the subjects a and b, and to make any appropriate recommendations.

Having accepted the invitation, a joint group of some 15 persons was assembled to carry out the task. The study, while clearly carried out on a crash basis (less than one year), nevertheless, developed a number of important recommendations.[106] For one, it was the unanimous opinion of the 11 countries reporting to the committee on the subject that a universal recording was not a feasible method for determining dose to the population and the report listed a number of the reasons for this. Instead it was agreed that satisfactory methods have been developed by a specialized survey organization for selecting a representative sample of the population and for making personal contact with this sample by trained surveyors; to collect the required information proven techniques can be applied for tallying the final results and determining their probable errors. It is interesting to note that this technique has been used successfully within the past decade by the U. S. Public Health Service.

Advice for the sampling procedure and the kind of information to be obtained was outlined in considerable detail.

At the outset of the discussions, it was decided that the study group should consider or develop methods of arriving at the genetically significant annual gonad dose G_m of the population for medical exposure; this was recognized as probably the most important immediate objective. Thus, for purposes of the report, it was concluded that the "genetically significant annual gonad dose G of a population is defined as the summation of the product of the average annual gonad dose D_i in millirads received by each person in age group i multiplied by the average child expectance P_i of the age group, multiplied by the number of individuals N_i in the age group, divided by the expected number of offspring of the population."

While the invitation from the UNSCEAR did not ask for suggestions on methods of reducing radiation dose to the population, the group felt that this was a fertile field which warranted treatment and, in the appendix, made a number of suggestions to this end. The appendices also included summaries of existing data on gonadal dose arising from medical procedures. This was rather spotty and clearly better and more detailed in some countries than others.

The report made a number of recommendations:

A. That the basic studies be continued and extended making use of suitable ionization dosimeters to obtain data for the preparation of standard tables for gonad dose.

B. All countries make evaluations of the genetically significant dose through the analysis of film records and if a dose so calculated exceeded a few percent of natural background to undertake a more detailed analysis.

C. Where required, the more detailed analysis should be obtained by means of sampling programs.

D. Prior to initiating sampling programs, a number of pre-surveys should be conducted, and

E. In preparation for the main sampling program, careful planning should be effected.

Members of the Study Group were asked to urge their respective countries to set up new survey programs or improve the existing ones as needed. This was done in a number of countries and when the final study was completed and reported in 1961, most of the countries except the United States had indeed provided substantial useful data. Attempts to interest and involve the U. S. Public Health Service produced no results for the International Report. However, for other reasons and with other pressures, they did take up the study soon after and carried out a very successful and useful survey.[41]

The second report of the study group, issued in October 1961,[110] was designed primarily to elaborate on numerous points and recommendations raised in the earlier report. In addition, while the earlier report emphasized the genetic effects, the 1961 report dealt more with somatic effects. No new basic recommendations were developed. For example, illustrative cases were presented for the purpose of showing how the biological and physical parameters could be applied to specific problems in practice. This was then followed by a number of examples that would be regarded as of a general practical nature. The report is much too detailed to be summarized in this writing.

It is interesting to note that the ICRP and ICRU, for the first time, received some moderately substantial financial assistance for the purpose of carrying out their meetings and paying travel expenses, when necessary, for the participants. During this period most countries were very generous in travel allowances for their scientists; however, by the end of the next decade the reverse situation prevailed so that not only were the parent organizations of workers not willing or able to pay for participation in such international activities, but also the international organizations themselves found it increasingly difficult to obtain financial support whether from other international agencies or foundations.

PUBLIC HEALTH SERVICE - 1952

It may be worth noting a few details of the early role of the U. S. Public Health Service in the radiation protection picture. In the early 1950's there were only a few individuals in the Commissioned Corps of the Public Health Service who had some background and training in radiation matters and who felt some concern for the lack of a defined role in the radiation protection field on the part of the Public Health Service. Probably through their efforts the Surgeon General at the time put out a trial balloon proposal suggesting some degree of radiation control by the Public Health Service. Admittedly, fearing a negative reaction and opposition from the medical profession, control of their activities was specifically excepted. When this proposal and the exception came to the attention of the American College of Radiology, they felt that if there was to be any control problem it certainly should include the medical profession as well as others. To make this point, in early 1952, they sent a small delegation consisting of Dr. Eugene Pendergrass, G. Failla, W. E. Chamberlain, and L. S. Taylor to Surgeon General, L. Scheele, to urge the start of at least an exploratory program of radiation exposure regulation and to have this program include medical radiology. No specific action was taken, but equipment inspection programs were instigated in Public Health Service installations. Also, a cooperative project was undertaken with the NCRP, resulting in the issuance of a report on radiation regulation (see Report No. 19).[73]

Dr. Lee Burney succeeded the retiring Surgeon General and in January 1956 Taylor gave him and Dr. J. Porterfield some background on the earlier discussions with PHS and urged that he consider the possibility of establishing some kind of active radiation program in the service. As a result of these conversations, and the generally changing "radiation climate," a conference, including most of the bureau and department heads and their deputies, was held at the Department of Health, Education, and Welfare. The conference was addressed by L. S. Taylor primarily on matters of radiation protection, and by Dr. H. O. Wyckoff on matters of radiation measurement.

CONGRESSIONAL CONCERN OVER RADIATION HAZARDS

Beginning in 1957, a new and important dimension was added to the consideration of problems of radiation hazards and protection against them. This new dimension was Congressional concern and interest.

Because of the national and worldwide reaction to the detonation of nuclear weapons, and because of the growing amount of inaccurate and provocative reporting in magazine articles and books that were appearing on the subject, and because the success of our Atomic Energy Program was considered an essential part of our defense system, the Joint Committee on Atomic Energy began a series of hearings to develop information on radiation hazards and associated problems. The first of these hearings carried out during late May and early June 1957, brought out to the committee and to the public so many new facets of the radiation exposure problem that the committee very wisely decided to step up the frequency of the hearings and the breadth of their coverage.

At the outset there was one hearing after another in close succession and they continued on into the 1960's on an intermittent basis (A list of the hearings will be in the bibliography, and with each will be included specific reference to the phases of the hearings presented by or closely related to the NCRP.) It might be noted that from the very outset it was recognized by the Joint Committee that the NCRP was essentially the only national body having broad and long-time experience in matters of radiation protection and that the ICRP was the only international organization having similar experience.

No attempt will be made to discuss the hearing reports in detail. They are much too complex and the record includes many questions that are approached from different angles and with different social, political or economic or other interests. It is the writer's opinion that the collection of these reports from the Joint Committee probably represents one of the finest contemporary collections of radiation protection information that exists. There is much material in these reports which has never been published elsewhere and while it may be regarded as unreviewed, it is important as an indicator of the nature of the problem and what some of the solutions might be.

The broad findings of the various hearings, especially the earlier ones, have proven to have had great influence on subsequent trends in radiation protection standards and their enforcement at various levels of government. They were important also in opening up to the public, almost for the first time, many of the problems and ramifications of radiation protection. Because of the generally inaccurate or inept manner in which this material was presented through the public press, the hearings had a tendency to generate unwarranted alarm and to instigate activities in some of the government agencies that were either not needed or were greatly extended beyond their period of usefulness.

The first set of hearings on "The Nature of Radioactive Fallout and Its Effects on Man" were held from May 27 to June 3, 1957:[144–147] they contained a number of surprises which clearly had influence on subsequent hearings. Perhaps the most important of these was the discovery by the Joint Committee and by scientists and the general public, who were not experienced in radiation protection problems, that it was not possible to set exact standards for radiation protection. Apparently there had been a general belief that it was possible, on the basis of scientific knowledge, to set some level of exposure above which radiation injury would result and below which it would not. In fact, there was apparently considerable surprise generally that this same situation existed for many other agents regarded as toxic or deleterious to health.

A corollary to this was that the term "safe" or "unsafe" had only relative meanings in the radiation area. Apparently, it took considerable time for this point to sink home to the press — if it has yet — because a great deal of public attention was devoted to the fact that scientists could not agree among themselves on these radiation matters. It did no good that the problem had been outlined in detail in the 1949 NCRP Report No.17, and that the general risk philosophy had been outlined in that report and many others since.[42,113,71]

Another discovery was that while radiation from fallout was by no means unimportant, its magnitude in areas where any appreciable segment of the public might be exposed was clearly less than expected. It was also clear that because of the possible ingestion of longer lived radionuclides, the implications for long time-exposure of members of the population might have serious hazard implications if it occurred on a widespread basis. Attention was also directed to other aspects of the nuclear energy program centering about the exposure of radiation workers. Here it was brought out that because of the extreme caution with which the Atomic Energy Commission programs had been conducted, there was very little experience with radiation injuries to fall back upon, or from which information on biomedical effects at low doses might be better evaluated.

Throughout the course of the hearings, there was extensive interrogation of various government agencies, especially the Atomic Energy Commission, regarding the radiation standards that they employ. It was something of a shock then to discover that without exception they kept referring back to those of the NCRP, or the ICRP in some instances. The realization gradually developed that all of the radiation protection activities for which the Federal Government was spending large sums of money, and for which it had the basic responsibility for safety under the Atomic Energy Act of 1954, were, nevertheless, in the hands of a non-governmental organization over which the government had no control. There was an immediate clamor to do something about this.

The fact that the government had no control over the development of its protection standards, was not unique. It was very much the pattern of the standard-setting philosophy in this country, that standards would often be developed by private bodies representing the private sector, but always insuring that there was input from the government reflecting its special needs. This is in accord with much of the philosophy in a free country whereby the government does not very often set the standards by which it controls its own activities. This is not to say that the government had no role whatever in setting standards; it does indeed have a strong role, but in most areas the role has been one of collaboration rather than one of domination.

FEDERAL RADIATION COUNCIL (FRC)

Because of the uncertainty of government influence over radiation protection standards, an investigation or study of the problem was undertaken by the Bureau of the Budget under Mr. Robert Cutler to determine just what the government should do about it. Perhaps one of the more important findings that developed during this study was the considerable amount of interagency rivalry or jealousy regarding radiation needs. Another much more to be expected, of course, was that the different agencies would have quite different goals that they would want to achieve in the uses of radiation, and these could strongly influence their attitudes towards any particular or general standards.

For example, it could have been the interest of the Atomic Energy Commision to promote the uses of nuclear energy both for purposes of defense and for medicine, industry, and research. On the basis of their own interests, it might be expected that they would want to promote a rather liberal standard. Then there was the Department of Defense for whom the prime interest was radiation involved in weapons for military applications. It would not have been unreasonable to have expected them to have been extremely liberal in their demands. The Department of Labor might have wanted to be stricter because of the involvement of people in radiation work. They and the Public Health Service could well have liked even stricter standards — and so it went. It is not that these agencies clearly displayed these large ranges of opinions, but it was clear to Cutler that if any one of them had a prime responsibility for setting of the radiation standards, those were the biases that the Bureau of the Budget would expect to find and which would be expected to cause difficulties with other agencies.

Out of the Federal organizations working with radiation, the one having the most activity, outside of the Atomic Energy Commission, was probably the National Bureau of Standards which had long been responsible for the development of radiation protection information and radiation measurement standards, and which had been the informal home of the NCRP since its establishment in 1929. There were indeed some tentative discussions with the Bureau of Standards relative to the possibility of their taking on the protection standards problem for the government. This idea was abandoned mainly because the NBS had no statutory authority or responsibility in the regulatory field.

Another possibility explored was that of "federalizing," in some way, the NCRP so as to make it a government controlled and operated committee. This, however, was objected to by the committee itself and so some thought was given to the possibility of organizing another committee of some kind in parallel to the NCRP. This was given up also.

It was out of this welter of confusion that the concept of the Federal Radiation Council was conceived. Since there was no rationale by which a single department could be given the radiation protection responsibility, and since experience with interagency committees indicated that they would probably be too cumbersome to achieve the desired objectives within reasonable times, the responsibility was finally given directly to the President. He in turn, had to be properly advised through some mechanism, and so the Federal Radiation Council was established. Its membership consited of the Secretaries of Commerce and Health, Education, and Welfare, Departments of Defense and Labor, Chairman of the Atomic Energy Commision, and the President's Science Advisor. Other agencies such as Agriculture, Interior, State, etc. were added later.

Of course, it was not reasonable to expect the agency heads to be familiar in detail with the important aspects of radiation protection. To provide them with suitable backup, what was known as a "working group" was established and consisted of one or more representatives from each of the agency members of the Council. These representatives were to be selected within the agencies from whatever organizational level it was necessary to reach in order to provide the necessary technical competence. It was the wish of the government not to take any steps which would destroy the independence and functioning of the NCRP. Consequently, there were frequent discussions between Cutler and the chairman of the NCRP and it was agreed that the final order setting up the Federal Radiation Council would not be made until it had been cleared with the NCRP; this was not done at the instigation of the NCRP, but probably as a token of confidence on the part of the Bureau of the Budget. In any case, in early August 1959, a proposed statement for an Executive Order was read to, and approved by, the NCRP and then issued.

Issuance of the Executive Order was followed very closely by an amendment of the Atomic Energy Act of 1954, which followed essentially the order but added the requirement that the Federal Radiation Council should consult with the President of the National Academy of Sciences and the Chairman of the National Committee on Radiation Protection. This was formalized in Public Law 86-373.

NCRP RELATIONS WITH NBS

Still another outgrowth of the 1957 Hearings and the unusual role of the NCRP in government radiation protection standards was the effect upon the Director of the National Bureau of Standards and his attitude towards the NCRP which the Bureau had been fostering in an unofficial way for some 30 years. There was nothing irregular about these relationships between the NBS and the NCRP; this kind of support to outside organizations had long been encouraged as a part of its public responsibility. However, there had not been any cases before where it appeared as though they might be fostering an organization which could be in the position of developing recommendations and philosophies in the radiation protection field that might be contrary to those used by the Federal Government as a whole. While the possibility of this was recognized as being fairly unlikely, nevertheless it could not be written off entirely.

In addition, there was a degree of laxity in the NBS-NCRP committee relationship in that it did not really fully comply with the 1955 government policy on committees. This policy in essence required that the committee be named by whatever agency was involved and that the members be cleared by that agency and invited to participate in the committee's activity. Also the agency was required to designate the chairman. As it turned out, none of these particular features was being met by the NCRP. They had been discussed previously with the legal staff at the Department of Commerce and there was an understanding that the relationship could continue as it had started. However, when certain writers in the public press and certain elements of organized labor began making individual attacks upon the NCRP, its mode of operation, its chairman, etc., it was clear that something had to be done to avoid embarrassment to the NBS.

In 1957, a meeting of the full NCRP was held at the NBS at which time the Director attended and took part in many of the dicussions.

No effort was made by the Director immediately to dissolve or alter the relationship with the NCRP. However, he did recommend that the Committee itself, examine its situation and study the question of making a clear separation from the NBS. He further recommended that the committee incorporate in some manner, or preferably seek a Federal Charter through the Congress of the United States.

There was, however, one other important outgrowth from this seeming conflict between the NCRP and its foster parent. Over the years the Bureau of Standards had, as a part of its public policy, published the reports of the NCRP as a part of its normal Handbook Series. This, of course, was a great aid in disseminating its recommendations. On the other hand, in spite of all of the disclaimers that were made — and they were made frequently — there were many people in the public who because of the government imprint on the cover of the report, assumed that the report represented government recommendations rather than those of an independent private body having no legal authority. This apparently had not disturbed anybody very greatly because recommendations had come to be recognized as objective and well considered. Up until about the time of the JCAE hearings the reports of the committee has been pretty much physical in nature and therefore could be readily regarded as coming within the normal competence and areas of concern to the Bureau of Standards. However, at just about this time the NCRP was preparing to publish its revision of the earlier report on "Maximum Permissible Body Burdens and Maximum Permissible Concentration of Radionuclides in Air and in Water for Occupational Exposure." This was an activity that had continued to be carried out under the Chairmanship of Dr. K. Z. Morgan and, as already noted above, the same work was being carried out by him for the ICRP.

Because this new report was going to contain an extensive discussion on biomedical matters and considerations of biological nature, the Director of the Bureau was concerned because these fields were not within the Bureau's competence and, at first, refused to allow the report to be published in the Handbook Series.

Finally, as a compromise, it was agreed to publish the formulation and some of the tabular material of the report which contained the essence of the conclusions, but to omit all of the biological and biomedical discussions, and let this be published as a part of the ICRP report. In addition, it was insisted that inasmuch as the recommendations might impinge upon the statutory responsibilities of both the U.S. Public Health Service and the U.S. Atomic Energy Commission, it would be necessary to determine that these agencies would not object to the publication of the recommendations by the National Bureau of Standards. Such assurances were obtained although they involved no commitment on the part of those agencies to adopt the recommendations. Nor was the publication to be construed as a recommendation by the National Bureau of Standards for adoption inasmuch as the important medical and biological factors involved in developing the recommendations were clearly outside the Bureau's area of technical competence.

The best that could be done for reference to the important biomedical sections was to list the contents of the ICRP Internal Radiation Report. The whole series of compromises was considered to be unsatisfactory at best and spurred the efforts of the NCRP to establish its independence by some appropriate means. (For more detailed discussion of this problem, see the preface to NCRP Report No. 22, NBS HB69, 5 June, 1959.)[78]

PUBLIC HEALTH SERVICE ACTIVITIES

Another outgrowth of the 1957 Hearings of the JCAE was sharply increased sensitivity on the part of the U.S. Public Health Service to its responsibilities in the field. As already noted, over the preceding few years they had been slow to pick up their responsibilities in this area. While the problems were clearly recognized at the staff level, it did not seem possible to adequately impress the top levels as to the radiological needs. The hearing served to accomplish this because the Public Health Service had almost no role to play in them. Possibly because of this, or it might have been developing in any case, Dr. Burney, who had become the Surgeon General in 1956, was very alert to, and concerned about, the problems of public health responsibilities in the radiation field. In 1958, he called together a group of professional experts in the field of radiation hazards and organized what was known as the National Advisory Committee on Radiation (NACOR). (It was once thought that this particular name, so close to that of the National Committee on Radiation Protection, was chosen so as to overshadow the independent committee. However, this was not the case and it was customary for the department to name all of its committees, "National Advisory...") The committee operated under the Chairmanship of Dr. Russell Morgan, who was a member of the Commission Corps Reserve of the Public Health Service — and also a member of the NCRP. He thus qualified as a suitable chairman under the government rules requiring that the chairman be a government employee. In addition to the chairman, the Surgeon General selected the members of NACOR for three-year terms. It was an effective body and during the eight or nine years of its existence it was probably one of the most important elements in getting the Public Health Service thoroughly involved in radiation matters.

The NACOR reports will be discussed later in connection with the other Public Health Service activity.

NATIONAL ACADEMY OF SCIENCES

Mention has already been made of the studies made by the National Academy of Sciences by their Committees on the Biological Effects of Atomic Radiation. It was their report on the Genetic Effects of Atomic Radiation issued in June 1956,[175,176] that led to a lowering of the basic permissible dose for radiation workers. In 1960, the committee reviewed the genetic situation and issued a supplementary report.[181] This recognized some later findings, especially those relative to the effects per unit of radiation dose received at low dose rate, which they considered to probably be less than previously estimated. This was an important change and has been verified and extended by later work as well as from early data obtained at high dose rates. The committee, however, did not feel that this new information would warrant changes in their earlier recommendations with regard to the limitation of average gonadal dose to the general population during the first 30 years.

In 1956, and continuing thereafter, at least through 1960, there were several other committees, beside that on genetic effects, which were set up by the Academy as a part of the BEAR group. These were as follows:

1. Pathologic Effects of Atomic Radiation, Shields Warren, Chairman. This committee had several subcommittees: (a) Acute and long term hemotological effects,[184] (b) Internal emitters,[185] (c) Inhalation hazards,[182] and (d)

Permanent and delayed biological effects of ionizing radiations for external sources.[183]

2. Meteorological Aspects of the Effects of Atomic Radiation, Harry Wexler, Chairman.

3. Effects of Atomic Radiation on Agriculture and Food Supplies, Dr. A. G. Norman, Chairman.

4. Disposal and Dispersal of Radioactive Waste, Abel Wolman, Chairman.

5. Oceanography and Fisheries, Roger Revelle, Chairman.

The output of these committees provided needed and valuable information in the establishment of Radiation Protection Standards.

FEDERAL RADIATION COUNCIL

The Federal Radiation Council was organized soon after issuance of the Executive Order and Public Law establishing it. As already noted, one of its first actions was to establish the principle of a working group made up of individuals professionally familiar with the subject, and also at high enough organizational level to evaluate policy considerations. The group met at least once a week and sometimes more frequently. Its first Chairman was Dr. A. V. Astin, then Director of the Bureau of Standards. L. S. Taylor was allowed to participate as an advisor to Astin when needed, but not as a member of the working group. The reason for this was because of the "possibility" of there being a conflict of interest in his being a member of the independent NCRP and yet sitting with an official government committee. Dr. Astin's Chairmanship was continued through the issuance of the initial report. Dr. Donald Chadwick, from the Public Health Service, was Secretary of the Federal Research Council (FRC) from its initial formation until 1963 when the position of Executive Director was established and filled by Dr. Paul Tompkins.

There were still difficulties in operating any such arrangement because all of the members of the working group were individuals with additional and primary management responsibilities and could not spend the kind of time that was being demanded of them in the new activity. The first report of the Council, which was issued in 1960,[128] was prepared largely by bringing together a small corps of experts from various government agencies or their contractors into a temporary staff which operated full time for several months. Their reviews of the background material, published at that time and in another report about a year later, were excellent summaries of the situation and delineations of the problem, but did not develop anything basically very new.

It is significant that the special staff, followed by the regular working group, and with a considerable amount of outside consulting, arrived at a set of basic radiation protection standards which were the same as those recommended by the NCRP. This agreement should not be ascribed to the fact that there was so much overlap in professional membership between the various groups. Actually, the problem was treated with a high degree of objectivity and a special effort was made to obtain the views of people who had not been previously involved in the development of protection recommendations. But after all, any protection recommendations that were to be developed must have been based on available biomedical and other technical information. If everyone is familiar with all of that information, it is not surprising that the final results should have been pretty much the same.

One other item occurred early in the discussions both with the Federal Radiation Council itself and with the temporary staff and working group. It was explained to these groups how some of the NCRP recommendations had been arrived at, how attempts were made to build some flexibility into their utilization, and how, since the basic biomedical information was not known with great accuracy in the first place, substantial "safety factors" had been built into the discussions. Safety factors in the more usual sense have not been applied to the derivation of numerical recommendations by the NCRP. One factor might be based on the lowest dose below which there is observable damage (toxicological approach); the other would be based on an over-estimate of damage expected on the basis of a linear, no-threshold relationship between effect and dose. Maybe both apply, e.g., the first in the radium case and the second in the dose-equivalent cases. The point was repeatedly made that the NCRP tended to look upon its own recommendations as a form of guidance leading to good radiation protection practices, and it was concerned that the use of their numerical quantities in government regulations would only accent the rigidity.

To overcome this, or perhaps work towards more flexibility, the Federal Radiation Council adopted a new term, "guide," in place of "standards." A radiation protection guide (RPG) was defined as "the radiation dose which should not be exceeded without careful consideration of the reasons for doing so; every effort should be made to encourage the maintenance of radiation doses as far below this guide as practicable." This seemed a useful concept and, of course, was in complete accord with the general philosophy that had always been used by the NCRP. However, it has not succeeded in totally avoiding the problem of inflexibility. A regulatory authority simply cannot live with flexibility. Therefore, regardless of whether it is called a guide or a standard, as far as the regulator is concerned whatever number is set, it is still in his mind a boundary between compliance or non-compliance. To that extent the use of a new term has failed of its intention. On the other hand, as will be pointed out later, it has served a useful purpose in dealing with some of the problems, for example, in the institution of countermeasures in the event of local increasing levels of radioactive fallout due to an accidental release of radioactivity. In this situation they are able to deal with ranges of exposures during which certain actions are taken, and other ranges in which other actions are taken; here the guides are used to mark the borderlines between the ranges and in this application have been useful.

In 1964, the Federal Radiation Council introduced an important new concept, that of the Protective Action Guide (PAG).[135] The PAG was defined as the projected absorbed dose to individuals in the general population which warrants protective action following a contaminating event where the projected dose is that dose which would be received in the future by individuals in the population from the contaminating event if no protective action were taken. (As an example, a projected dose of 30 rads to the thyroid of individuals in the general population was recommended as the protective action guide for iodine-131. As an operational technique it was assumed that this condition would be met effectively if the average projected dose to a suitable sample of the population did not exceed 10 rads.)

In the application of the protective action guides the following guidance was provided:

1. "If the projected dose exceeds the PAG, protective action is indicated.

2. The amount of effort that properly may be given to protective action will increase as the projected dose increases.

3. The objective of any action is to achieve a substantial reduction of dose that would otherwise occur — not to limit it to some prespecified level.

4. Proposed protective actions must be weighed against their total impact. Each situation should be evaluated individually. As the projected doses become less, the value of the protective actions becomes correspondently less."[135]

Protective Action Guides were also developed for strontium-89, strontium-90. and cesium-137.[139]

AD HOC REPORT ON SOMATIC EFFECTS OF RADIATION (NCRP)

The preceding discussions have carried the development of protection philosophy through a series of basic changes. During the same period, there were equally important developments in radiation protection management and the development of new attitudes towards legislation and regulation. At the same time, there was not only the development of a public consciousness about radiation matters but also there was strong reaction at the many levels of government and the many scientific levels about the problems raised by radiation uses.

The prime element leading to the downward revision of the basic permissible dose for radiation workers and the introduction of the dose limit for population groups in 1957 also raised the question as to the adequacy of the philosophy with regard to the somatic effects of radiation, especially on population groups. It was clear in thinking about this problem that it involved not only the extent of our knowledge about the somatic effects of small doses of radiation, widely scattered among the population, but also brought us clearly face to face with the concepts of the risk philosophy which was first expressed by the NCRP in its 1949 report (NCRP Report 17).[71]

To meet this objective, the Executive Committee of the NCRP, at its meeting in November 1958, undertook to reexamine the problem of exposure of the population to man-made radiations from the point of view of somatic effects as distinct from genetic effects. This was done even though there was no awareness of any new basic information on somatic effects of radiation upon which it could with sound reason recommend specific changes in permissible exposures for individuals or population groups. It thus appeared desirable to make a new and independent examination of the problem for the purpose of affirming or modifying the views of the NCRP. For this purpose, a special Ad Hoc Committee under the Chairmanship of Dr. Hymer Friedell was organized and asked to examine the question further.

The committee was organized to include some members of the NCRP who had participated actively in the development of its basic criteria, together with some others, who had had little or no connection with the NCRP. The report had very important implications and set clearly the assumptions or postulates which it felt should be used as a basis for future considerations of radiation protection philosophy. When completed, it was referred to the Executive Committee and Committee 1 for further consideration, and for the possible formulation of specific values to be recommended as maximum permissible dose.[79] As events have proven, the general issues raised by the Ad Hoc Committee have become a basic part of the philosophy and thinking of the NCRP as well as all other protection bodies. Since the report contained recommendations that were more of a philosophical than practical nature, the NCRP did not publish it as specific recommendations but as a public statement indicating the directions in which some of its thinking would probably move.

The committee had started with an examination of the dose-effect relationship at low doses and concluded that the data did not permit a distinction between possible linear or curvilinear dose effect relationships at low doses, or determine whether or not there might be thresholds. They further concluded that there was insufficient knowledge of the mechanism to serve as a guide in areas where actual data were not available. Their conclusion followed that it was prudent to be conservative and to use assumptions which, if in error, would be likely to overestimate the effect of low doses rather than to underestimate them and to this end adopted the postulation that a proportional relationship between dose and effect existed over all dose ranges.

Having decided this, the Ad Hoc Committee addressed itself to the question as to whether the average population dose should differ from that for occupationally exposed groups. Their conclusion was that it should be substantially less for the following reasons:

1. The general population is much larger and if exposed to the same dosage, there would be the risk of the correspondingly larger number of individuals with injurious effects.

2. Employment involving occupational hazard to exposure is voluntary, and the extent and nature of the exposure can, in principle, be foreseen by the individual accepting any risk that may be involved.

3. Industrial workers were relatively carefully screened.

4. It was possible in industry to carry out specific evaluation and control of radiation hazards by means of monitoring and other studies.

5. Children and embryos might be particularly sensitive and could be generally excluded from groups receiving the maximum permissible occupational dose.

6. The number of years of exposure to radiation for occupational reasons would be much less than the number of years of exposure to environmental sources of radiation, and

7. If industrial hazards do exist, it is obvious that any of these hazards, one of which is radiation, should not be spread beyond the individuals in that particular occupation. To do otherwise, the risk to the total population could be unacceptably high because of the contributions from all kinds of occupational hazards in the society as a whole.

On the basis of the general principles outlined during the meetings of the Ad Hoc Committee, and touched on briefly above, the following recommendations were made for the guidance of those concerned with the establishment of tolerable somatic levels from widespread radiation in the environment.

1. Present evidence, not being sufficient to establish the dose response curves for somatic effects at low doses, and in the absence of such information, it was *postulated* that there is a proportional relationship between dose and effect and that the effect is independent of dose rate or dose fractionation.

2. On the above postulate, or any other regarding dose effect relationships, it follows that even the smallest dose might be associated with some risk. Under such circumstances, the exposure

of the population to any increase in radiation should not occur *unless there is reason to expect some compensatory benefits.*

3. Because of the limited information, it was not possible to make an accurate estimate of hazards or the benefits of a specific level of radiation. Therefore, pending more precise information, it was recommended that the population dose limit for man-made radiation be based on, or related to, the average natural background level.

Recognizing that it was not their responsibility to determine an exact level, the belief was expressed that the population dose limit from man-made radiations, excluding medical and dental sources, should not be substantially higher than that due to natural background radiation, without a careful examination of the reasons for, and the expected benefits to society from, a larger dose. This applied to individuals.

The further opinion was expressed that because of fluctuations in time and location, the average dose to the population will be considerably less than the dose limit.

4. For purposes of computation, it should be allowable to average the doses over a suitably long period of time, e.g., one year, and a reasonable sized population.

5. For radiation sources, such as radioactive strontium and iodine, which deliver radiation predominantly to one organ or tissue, the maximum permissible dose, or dose limits, should be established on the basis of the anticipated effects in that tissue or organ.

6. It was recognized that it was not possible to monitor the population dose solely by measuring the dose to the individuals and, furthermore, that any effective control over radiation levels, must be directed at the levels of radiation and radioactive materials in the environment. This meant that maximum permissible levels or dose limits would need to be established for such factors as food, water, and air and the levels so set that the typical person in any given area would not receive more than the established allowable dose when all sources are combined.

7. It was recognized that the establishment of any environmental levels would involve assumptions and conversion factors to translate these into human body levels. These factors may be expected to change with new information so that environmental levels might be expected to require continuous revision, even though the permissible amounts to the body are not changed.

8. Recommendations leading to the establishment of permissible levels for medical and dental exposures to the patient could not be given because, for somatic effects of radiation, the possible harm and the prospective benefits occur to the same individual in contrast to radiation involving the genetic materials. The committee urged that continual caution be experienced to maintain radiation for medical and dental purposes at the lowest feasible level.

9. Finally, the committee emphasized that under one of the primary postulates that they made in their report (non-threshold linear dose-response) the biological effect does not suddenly change from harmless to harmful if any particular permissible dose is exceeded. Any permissible level which may be chosen is essentially arbitrary and every effort should be made to keep the radiation doses as far below a permissible level as possible. On the assumptions noted above, any radiation dose should be thought of as being tolerated only to obtain compensatory benefits.

In retrospect, the prudence and wisdom of this report have been amply verified over the succeeding decade. Its recommendations have indeed been used as the basis for the 1971 Report of the NCRP on "Basic Radiation Protection Criteria," completed after nearly ten years of continuous study and evaluation of available information.

NCRP Reports Up To 1961

During the period in which the development of changes and protection philosophies was active, various subcommittees of the NCRP continued with their broad assignments. Most of these studies involved applications of the basic principles, or the development of measurement procedures for evaluating radiation exposure problems under wide ranges of conditions.

Because measurement problems would play an important role in future NCRP activities, the committee's scope of interest was broadened to include four subcommittees devoted to measurement problems alone and the name of the Committee was broadened to the "National Committee on Radiation Protection and Measurement." (Since the symbol NCRP had become so widely known and accepted, the M was not added but in references to its work in the literature, it is frequently referred to as NCRPM.)

NCRP REPORT NO. 20

The Report on "Protection Against Neutron Radiation Up to 30-Million Volts," prepared under the Chairmanship of Dr. H. H. Rossi, was issued in November 1957.[75] (It was started initially under the Chairmanship of Dr. D. B. Cowie, who had to withdraw for reasons of health.)

While the recommendations of this report took into consideration the revised permissible dose levels issued by the NCRP in January 1957,[74] it left its requirements for the maximum permissible dose for radiation workers somewhat more stringent for neutrons; that was because of some of the uncertainties regarding the relative biological effect of neutrons and it was felt wise to be more, rather than less, cautious.

The scope of the report included neutron energies of up to 30 MeV. Although both theoretical and experimental information was sparce beyond about 10 MeV, the higher limit was chosen because many neutron generators at that time operated within the wider range. The comparatively few sources producing neutrons in excess of 30 MeV usually attain energies several times as great and a substantially different protection problem would be involved. The subject of neutron protection and reactors was limited to considerations arising in routine operations.

The report began with the discussion of the current status of physical and biological information, especially in relationship to neutrons, biological effects, relative biological effectiveness, etc. It also included a discussion of neutron detectors, the measurement of neutron flux, and dose measurement generally.

The second part dealt primarily with the design of protective installations and the operation of neutron sources giving data which would be useful for shielding purposes.

The report ended with the presentation of a series of rules for protection against neutron radiation from accelerators and reactors, and the general precautions involving surveys and health.

A considerable portion of the report was devoted to the inclusion of necessary data to achieve the recommendations given earlier. This included: (1) depth dose, (2) flux detectors and their calibration, (3) reactions employed in neutron production, (4) practical use of radiation instruments, (5) neutron capture gamma rays, (6) shielding calculations and data, and (7) neutron protection near high energy electron accelerators.

NCRP REPORT NOS. 21-28

In July 1958, a new report on the "Safe Handling of Bodies Containing Radioactive Isotopes," was completed under the Chairmanship of Dr. Edith H. Quimby.[76] This was essentially an updating of the previous report issued in 1953, following the same general format but taking advantage of new information and technology developed in the interim.

Report No. 22, updating the earlier one on "Maximum Permissible Body Burdens and Maximum Permissible Concentrations of Radionuclides in Air and Water for Occupational Exposure" has already been discussed in detail above.[78] Report No. 23, the first of the measurement subcommittee reports, "Measurement of Neutron Flux and Spectra for Physical and Biological Applications" was prepared under the Chairmanship of Dr. R. S. Caswell and issued in July 1960.[81] As its name implied, this report was devoted entirely to measurements and instrumentation problems and since it did not involve any basic changes in protection philosophy will not be discussed further.

Another of the measurement committees, under the Chairmanship of Dr. Victor P. Bond, prepared a report on the "Dose Effect Modifying Factors in Radiation Protection."[44] The primary objective of this subcommittee had been to examine questions of the relative biological effectiveness of radiation and they were urged to study the problem without inhibition, and to generate any new ideas which they thought might be useful in protection philosophy. This committee, originally organized under the Chairmanship of Dr. Wright Langham, spent over two years collecting all of the pertinent information relative to RBE with special emphasis on applications to protection.

In reviewing the experimental values available for relative biological effectiveness (RBE) and attempting to specify the quality correction factors (QF) for application to radiation protection, the subcommittee found that these values would be of limited usefulness if presented independently of the manner in which dose-effect modification factors were to be used in practical radiation protection. They thus set as a desirable objective the characterization of an individual's entire cumulative exposure status by a single number. To do this it was necessary to extend the present methods for adding exposure so that the incremental dose equivalent values could be for all types of external and internal irradiation. To achieve this goal they devised a new system for evaluating, summing, and recording occupational exposures as summarized below:

1. It was recommended that the risk from all radiation exposures be expressed relative to that from a basic whole body exposure to a standard low-LET radiation and, further, that cobalt-60 gamma radiation be designated as the standard low-LET radiation.

2. The risk from any exposure was related to that from whole body exposure to the standard radiation through an effectiveness factor, EF. A "correction factor" for external radiation was defined in their report as the product of several subfactors for radiation quality, body region irradiated and penetrating power. The product of these (dose in rad x EF), the dose equivalent, was expressed in terms of "exposure record units" (ERU, analogous to the term "rem").

3. It was also recommended that EF values be established for internal emitters, in order that the risk from known body burdens of internal emitters relative to that from external radiation could be representative in accumulative ERU record. Recommendations were given on how to accomplish this for representative EF values.

4. For all types of exposures, external and internal, the incremental dose equivalent values were summed.

5. Separate maximum permissible dose (MPD) values for organ or body regions were not required.

6. It was suggested that the MPD for external radiation be 6 ERU/qtr., subject to the cumulative rule of 5 (N-18) ERU, where N is the age in years.

7. Evaluation was considered to be made simpler because dose equivalent values in ERU derived from all types of exposure were summed, and total exposure could then be represented by a single value.

8. Experimental RBE data tended to support the QF values that are now in use and no change in the quality factor was recommended; however, as a result of including other subfactors, EF values for high LET radiations recommended by the subcommittee were frequently lower than currently employed QF values.

9. The basis for recommendations on partial body and penetration factors were given. The possibility of needing factors for additional variables, such as time patterns of irradiation exposure, was considered, but it was concluded that additional factors either need not be included or could not be provided at the time because of lack of appropriate data. An age factor was not considered necessary since age is adequately taken into account in the cumulative 5 (N-18) rule. (The age - 45 concept developed in NCRP Report No. 17 was still in being, although rarely used, at the time the Bond Report was written. It has been introduced in a different application in Report No. 37.[96]) It was clear that these recommendations would need a great deal of study and evaluation and while the NCRP considered the report imaginative and useful, it decided against issuing it in the form of formal recommendations, as for all other reports, for fear that it might be taken as an official position which the main committee was not yet ready to take. The report went into considerable detail in dealing with partial body and organ doses which was unnecessary when doses were well below the MPD. Under those circumstances film-badge readings would have been adequate. An additional reason for not wanting to express official action on these recommendations was the fact that after the NCRP Committee on RBE had been in operation for two or three years, a similar joint committee between the ICRP and the ICRU was established. There was a substantial overlap of membership between these joint committees and the NCRP Committee, and, in fact, on several occasions the two groups met together informally. The ICRP/ICRU Report will be discussed separately.

The next report on "Measurement of Absorbed Dose of Neutrons and Mixtures of Neutrons and Gamma Rays" was prepared by a Task Group under the Chairmanship of Dr. G. S. Hurst (NCRP Report No. 25).[83] In a sense, it was an extension of the earlier report on the measurement of neutron flux and spectra for physical and biological application. However, it marked an important step in the recommendations on radiation measurements and the practical application of measurement techniques to the determination of absorbed doses under wide ranges of conditions. It had only secondary or corollary interests as far as the development of radiation protection criteria was concerned and will not be described further at this time.

NCRP Report No. 26 on "Medical X-ray Protection up to 3-million Volts" was prepared under the direction of Dr. T. P. Eberhard and issued in February 1961.[84] This was essentially a third edition of the report of Medical X-ray Protection (Report No. 6) that was issued in March 1949. During the interim, there had been substantial trends toward the development of radiation protection regulations for medical uses of x-rays and there had also been introduced a number of improvements in the general technology. Thus, while the report had updated the technical information from the earlier reports, it also gave particular emphasis to some of the multifaceted characteristics of the recommendations. For example, the maximum output or the leakage of x-rays through the tube housing can be fairly readily determined using short exposure times. While an exposure limitation based on such a determination might be a perfectly legitimate requirement to impose on the manufacture of the equipment, there would be no justification for extrapolating such a determination to continuous 40-hour-a-week operation if it were known that the duty cycle of the tube was only 5% of this maximum value. The philosophy of dose limits, based upon the cumulative RBE doses, as then enunciated by the ICRP and NCRP, and other than the previously used short-term exposures, made these distinctions imperative. At the operational level also it was stressed that a collimating cone which is provided but left on the shelf does not meet the requirements of the recommendations.

It was also felt necessary to emphasize that the report was not a legal document and that its provisions were not written to be used in their entirety as official regulations. Those recommendations phrased with the word "shall" were to represent requirements necessary or essential to meet the currently accepted standards of protection, while those using the word "should" represented advisory recommendations that are to be applied when practicable. These concepts were originally defined in Report No. 6.[60] It was felt desirable that these latter be applied whenever and wherever they are practicable. It was also recognized that in some cases the equipment had not yet been developed to the point where meaningful recommendations could be applied universally. In others, the recommendations had been met by the manufacturers of current equipment but the committee believed that the risk involved in the judicious use of older equipment was not sufficiently great to justify the condemnation of thousands of otherwise satisfactory units. Generally, the manner of use of x-ray equipment was considered to be of more importance than its mechanical design.

There were other considerations which might have been quite practicable or even considered mandatory in a large, very busy installation but which would be quite unnecessary in an installation with a single machine and a very low workload. It was emphasized that persons who were charged with the formulation of radiation protection regulations, but who lacked practical experience in the field, should seek the advice of competent experts before establishing rigid rules based on the recommendations contained in the report.

The basic requirements in the report were essentially the same as those of its immediate predecessor; the tables were revised to conform with the current lower maximum permissible dose and the differences in the maximum permissible doses allowed in controlled areas and in their environs. The formats was revised to avoid repetition.

The philosophy as it was originally expressed by the Advisory Committee on X-ray and Radium Protection some 30 years previously, namely that, "unnecessary radiation exposure should always be avoided and all exposure held to a minimum compatible with practical clinical requirements," was enlarged and emphasized.

The next report (No. 27) dealt with the "Stopping Powers For Use With Cavity Chambers."[85] This was prepared under the direction of Dr. W. C. Roesch and issued in September 1961. Its primary purpose was to present a critical review of the literature concerning the stopping power ratio that is used in the interpretation of cavity ionization measurements in radiation dosimetry in application of the Bragg-Gray principle (the underlying ionization measurement techniques.)

The last of the measurement reports which had been started some years previously was a "Manual of Radioactivity Procedures," (Report No. 28), prepared under the Chairmanship of Dr. W. B. Mann and issued in November 1961.[86]

This was a comprehensive and detailed report divided essentially into three parts: (1) radioactivity standardization procedures, (2) measurement of radioactivity for clinical and biological purposes, (3) disposal of radioactive materials.

In addition to the detail included in the report, it included an excellent bibliography. The report continues to be reasonably in date and of substantial value.

While this report was of great value in the whole technology of radiation hazard control, it did not bear directly on the basic protection criteria and hence will not be discussed further at this time.

RADIATION FROM TELEVISION RECEIVERS

By the middle 1950's, large-screen television receivers had become a common household device. While initially, these were mostly black and white, accelerating voltages were moderately high and there was concern about the possibility of radiation escape from the face of the tube or from the general housing. In November 1959, the Executive Committee of the NCRP made a statement with regard to the maximum permissible dose from television receivers.[80]

The statement indicated that from a genetic point of view, even sources of minute radiation might be of significance especially if they affected a large number of people. Rays emitted by home television sets were therefore of interest because of the high percentage of the population involved. To insure that the television contribution to the population gonad dose would be only a small fraction of that due to natural background, the NCRP recommended that the exposure dose rate at any readily accessible point, 5 cm from the surface of any home television receiver, should not exceed 0.5 mR per hour under normal operating conditions.

Laboratory and field measurements had shown that with the maximum permissible exposure level, the television contribution to the gonad dose at the usual viewing distances would be considerably less that 5% of that, due to the average natural background radiation.[45] Most of the present television receivers already met this requirement with a high factor of safety. In general, therefore, no changes in shielding of existing sets would be required. However, the recommended exposure limit was intended to insure that future television receivers operating at higher voltages would not contribute significantly to the population gonad dose.[92] This subsequently proved to be the case for properly designed colored television but, as will be noted, a new type of difficulty resulting in leakage radiation came to light.

JCAE Hearings

Reference has already been made to the hearings held in May and June 1957 on the "Nature of Radioactive Fallout and its Effects on Man."[144-147] It was during these hearings that direct Congressional attention was first directed to radiation protection standards and some of the associated problems. Portions of those hearings related directly to the protection criteria standards and general philosophy will be outlined briefly below. The hearings themselves and the record of them is much too extensive to consider in detail, but it should be pointed out that they represent a prime source of background information for the whole problem.

The principal points brought out in the first hearings were as follows:

1. All nuclear explosion can be expected to produce some degree of radioactivity and since there is no such thing as an absolutely "clean" weapon, attention from a radiation protection point of view had to be directed to all weapons testing.

2. While there was far from complete agreement on what happens to radioactive debris produced in man's environment from weapons testing, there was considerable evidence to indicate that in no part of the atmosphere is fallout uniformly distributed and, therefore, that the effects of fallout on the world's population could not necessarily be expected to be uniform or completely predictable.

3. There was general agreement that any amount of radiation, no matter how small the dose, might increase the rate of genetic mutation in the population. On the other hand, there was difference of opinion as to whether very small doses of radiation could produce similarly increased incidence of somatic conditions, such as leukemia, bone cancer, or decrease in life expectancy in a population.

4. There was general agreement that there is some limit to the amount of radioactivity and hence to the amount of fission products that man could tolerate in his environment.

A number of unresolved questions also emerged from the hearings. These included:

1. To what extent do the biological processes of plants, animals and human beings — under normal conditions — exhibit a preference for or "discriminate" against strontium-90 and other potentially hazardous isotopes that are taken into the human body? It was suggested that sampling and metabolic studies under way would develop a better answer to this question.

2. Is there a "safe" minimum level of radiation or "threshold" below which there is no increase in the tolerance of such somatic conditions as leukemia or bone cancer, or no decrease in life expectancy in a population resulting from the radiation?

3. What is the genetic "doubling dose" of radiation to man?

4. Should a distinction be made between the absolute numbers of persons affected by fallout, and percentages relating these numbers to the total population of the world, i.e., can we accept deleterious effects on a relatively small percentage of the world's population when the number of individuals affected might run into the hundreds of thousands? Overall national policy and great moral issues are involved.

The conclusions and the questions listed above were discussed in detail in the body of the hearings and summarized in the Joint Committee Print (Aug. 1957).[147]

Additional important hearings were held in January, February, and July 1959, on the broad subject of "Industrial Radioactive Waste Disposal."[148-150] Important points brought out and conclusions reached are outlined below:

1. Radioactive waste management and disposal practices had not, at the time, resulted in any harmful effect on the public, its environment, or its resources.

2. Low level wastes have been dispersed to nature (air, ground, water), and with or without treatment as required, under careful control and management. The problem may be expected to increase as the nuclear power industry increases in size or if acceptable limits of radioactivity in the environment are further reduced.

3. The final disposal of high level wastes associated with chemical reprocessing of nuclear fuels represents an aspect of the problem that, while safely contained for the present and immediate future, has not yet been solved in the practical long-term engineering sense.

4. It will always be necessary to use the diluting power of the environment to some extent in handling low level waste.

5. Suggestions for final disposal of high level waster include: (a) conversion to solids by one of several methods, (b) storage of solids in selected geological strata with major emphasis on salt beds, (c) disposal of liquids in the geological strata such as deep wells or salt beds, (d) disposal of liquids or solids into the sea. At the time it seemed that the conversion to solids and storage of these in salt formation seemed to be most favorable, and the least favorable was disposal of high level wastes into the sea.

6. The long-term responsibilities associated with the protection of Public Health and Safety in natural resources must be borne by agencies of the public, probably at several levels of government, but primarily at the Federal level.

7. International aspects of the problem are important considerations, particularly in connection with disposal into the sea.

8. Initial costs associated with waste disposal, though large in absolute numbers appear to be a relatively small fraction of unit nuclear powered costs and are within the realm of economic practicability.

9. Medical exposure from x-rays constitute the major source of radiation to man, and the contribution from radioactive wastes at present, seem substantially less than that from worldwide fallout from nuclear explosions.

An NAS-NRC Report entitled "Radioactive Waste Disposal into Atlantic and Gulf Coastal Waters" had suggested 28 possible locations for disposal of low-level waste in "inshore areas," and in some cases less than 15 miles from shore and in water less than 50 fathoms deep.[179,188] Nevertheless, the AEC indicated that it would continue to follow the NCRP policy of disposing of radioactive wastes in depths of not less than 1000 fathoms, which in most cases would mean at sites at distances from 70 to 150 miles offshore.

The details of these hearings are contained in the full hearing record and summarized in the Joint Committee Print, dated August 1959.[150]

Further hearings on "Fallout from Nuclear

Weapons Tests" were held in May 1959,[154] at which time the Committee considered the major unresolved problems as noted above and as listed in the 1957 summary analysis.

The 1959 Hearings served to demonstrate that firmer support could be found for some of the statements of conclusion that were regarded as somewhat tentative in 1957, especially in relationship to the nonuniformity of worldwide fallout distribution and the discrimination factors for strontium versus calcium in plant and animal metabolism.

The 1959 Hearings also served to focus attention on some new points of controversy, such as the so-called hot-spot problem and the importance of short-lived isotopes in the fallout occurring far away from test sites. Perhaps most important, the question was raised as to how to evaluate measured levels of fallout in the environment and in man in terms of some sort of measuring standard of criteria. Central to the lack of understanding on the last point were differences of opinion with regard to the applicability, for this purpose, of the so-called maximum permissible dose (MPD) of radiation from internal or external sources, and the so-called maximum permissible concentration (MPC) of radioactive materials in the body, the air, or in water.[155]

The problems of Employee Radiation Hazards and Workman's Compensation was the subject of a series of hearings held in March 1959.[152] Areas of agreement included the following:

1. There is need for coordinated guidance on adequate standards for maximum permissible exposure to radiation and these standards should be uniform throughout the United States in order to avoid conflict and misunderstanding.

2. More information is needed with regard to the effects of ionizing radiation on man and especially to determine at which point exposures are not injurious, if at all.

3. There is need for adequate enforcement of radiation standards when they are adopted in codes or regulations.

4. There is need for centralized records of individual exposures.

5. There is need for greater public knowledge of radiation hazards and its methods of control among workers, the general public, communities, etc.

6. Compensation laws should include at least the following provisions: (a) that radiation diseases should be classified as occupational diseases and be compensated for by all compensation systems, (b) the time limit for filing claims should recognize the fact that disability from radiation exposure might not occur until long after the original exposure, (c) any diseases caused by radiation should have complete medical treatment during the course of disability, (d) compensation benefits should be paid on the basis of loss of earnings or the loss of earning capacity, and (e) the standards by which the causal relationship between exposure and subsequent disability should be established.

The major disagreements that were brought out were the substantial differences of opinion as to how the needs and objectives could best be met. There seem to be about equally strong advocacy for administration of the compensation laws by the Federal Government and by individual state governments. There were also differences of opinion as to whether AEC should extend its regulatory control to mining operations and whether the regulation of isotope users should be removed from the jurisdiction of the AEC, together with the licensing of reactor installations. In general, most of the differences of opinion among witnesses occurred more on the method of carrying out the desired objectives than in the objectives or the ends themselves.

An especially important suggestion was for the establishment of a Federal Board to make recommendations on radiation standards and their applications. Membership of the board would comprise representatives of responsible Government Agencies, including the AEC, Department of Health, Education, and Welfare, the Department of Commerce, and the Department of Labor. In addition to coordinating activities at the Federal level, the board would also provide guidance to the various state governments with regard to standards.

In connection with this, it was noted that an Executive Order had been issued by the President on August 14, 1959:[46] (a) creating a Federal Radiation Council to advise him on standards and other matters, (b) directing the Department of Health, Education and Welfare to intensify its efforts in radiological health and to assume primary responsibility for the collection and analysis of data on environmental radiation levels, and (c) directing the AEC to assume primary responsibility for preparing the states to assume certain regulatory duties.

On September 1, 1959, the Joint Committee on Atomic Energy approved legislation (S2568, S Report No. 870; HR 8755, H Report No. 1125) providing a statutory basis for the Federal Radiation Council in increasing the members of the council from four to five. The President's Executive Order had named the Secretary of Health, Education and Welfare, the Chairman of the Atomic Energy Commission, the Secretary of Defense, and the Secretary of Commerce as members of the Council. The above legislation introduced by the JCAE added the Secretary of Labor to the Council Membership and required the Council to consult with qualified scientists and experts in radiation matters, including the President of the National Academy of Sciences and the Chairman of the National Committee on radiation protection. The legislation was passed by the Senate and the House on September 11, 1959.

Full details of the hearings can be obtained from the completed record and the Joint Committee Print, dated September 1959.[153]

Of all of the Congressional hearings on radiation matters to date, probably the most important series relating to radiation protection standards were those held by the Joint Committee in May and June 1960, under the heading "Radiation Protection Criteria and Standards: Their Basis and Use."[159,160]

The Joint Committee had now had a full three years of experience gained through their hearings and many separate discussions. The problems brought up during the hearings were better identified and sharpened and there were specific goals to be achieved through the 1960 Hearings, rather than a groping as to what the problem was all about and what its magnitude might be. Clear areas of government concern as to its responsibility in radiation matters, could now be attacked with a minimum of lost motion.

Therefore, it was the intent of the 1960 Hearings to emphasize concepts and understanding of subject matter from a broad point of view rather than the strictly technical aspects. They constituted, therefore, something of a departure from the pattern of earlier hearings on related subjects. In addition to the hearing record itself, the hearings were preceded by an "Advance Committee Print," which represented contributions of importance to the literature on radiation protection, particularly in its broader social economic and policy aspect. It is one of the most valuable collections of papers dealing with both the technical and philosophical aspects of radiation protection and radiation protection standards as of that date.

It was also the intent of the hearings to emphasize more heavily than in earlier ones the organizational and administrative aspects of radiation protection in the government. The Federal Radiation Council had been organized and in operation for a little less than a year and its first report issued.[128] Of specific interest was the question of who develops radiation protection criteria and standards and for what purposes, and who adapts basic exposure recommendations for actual application in atomic energy activities, whether by the government itself or under license or other control.

Because, at this date, the NCRP had been in operation for over 30 years and had been responsible for the development of most of the protection philosophy in this country, and because the Federal Radiation Council was charged with the responsibility for radiation protection standards in the government, it was natural that throughout the hearings there was a great deal of attention directed to the past and future role of the NCRP and the FRC.

While there is far too much material of import in the hearing record to permit discussion in detail, some of the key points and contributing discussions will be given below:

1. The ultimate need for radiation protection criteria or standards was recognized as arising from two principal facts. (a) the development of the uses of atomic energy and other sources of ionizing radiation that will inevitably be accompanied by the exposure of persons to man-made ionizing radiation, and (b) enough is known about the biological effects of radiation on man to permit agreement on the working assumption that for protection purposes all such exposures, however small, may have an associated biological risk.

2. It was recognized that experience in many aspects of public health practice, having to do with man-made threats to the general public health, are likely to provide useful experience in the application to radiation protection and the development of suitable criteria and standards. (It might be noted here that in later years it began to

be generally appreciated that in many respects much more was known about radiation as a public health hazard than most other environmental agents common in man's life.)

3. While there was generally agreement on the broad purposes of radiation protection criteria and standards, there was more difficulty in reaching agreement as to whether the basic NRCP permissible radiation dose and radium body burden recommendations should be used as a basis for decisions, guidepoints, and action in the somewhat limited field of the peaceful uses of atomic energy.

4. There was agreement that there is a large body of applicable knowledge of the biological effect of radiation on animals and man and that the state of our knowledge of the biological effects of radiation compares favorably with that of other hazards to health. As a corollary to this there seemed to be acceptance that the best working hypothesis for protection purposes is one attaching some risks of harmful biological effect to every dose, no matter how low. *It was also recognized that this assumption is not unique for radiation.*

5. There were divergent viewpoints of how social and economic factors had been taken into account by the NCRP in arriving at its basic permissible dose recommendations.

6. There was some uncertainty as to the scope of application for which either the NCRP or FRC radiation protection recommendations was meant; for example, there was contradictory testimony on whether the NCRP recommendations would apply to fallout.

7. Report No. 1, issued by the Federal Radiation Council,[128] was considered highly significant and marked a major turning point in the government's approach to problems and policies in the field of radiation protection. There was, however, considerable difficulty in understanding the scope and meaning of the FRC radiation protection guides as they might apply to actual practical conditions.

A number of major unresolved questions concerning the Federal Radiation Council were identified, including the following:

1. Will the President, with the advice and assistance of the FRC, make the basic judgments and exercise final and direct responsibility for setting the radiation protection standards, or will each Federal Agency perform this role in formulating its own standards and devising its own regulations?

2. What role will the FRC play in, and how will the FRC meet, the problem of ever increasing man-made radiation exposure levels? Will a system of allocation be followed or will each agency have the full limits for its own use?

3. How will the National Committee on Radiation Protection be affected in view of the role now being assumed by the FRC?

4. Do the activities of the FRC remove standard-setting functions from the application of usual administrative and procedural safeguards, such as the right of appeal, set forth by statute?

5. Does the FRC provide for adequate representation of Agency as well as public interest, and for adequate representation of professional, including scientific, disciplines?

6. To what extent should FRC be primarily a policy group, emphasizing broad social and economic and policy factors, or to what extent should it be a group devoted to the technical details of standard setting?

7. What does "normal peacetime operations" as used in the FRC Memorandum No. 1 mean?

8. A major unresolved question still exists as to how to keep the flexibility, characteristic of such guidance as the FRC Radiation Protection Guides, and at the same time meet the need for legal precision in regulatory applications.

9. There seemed to be a general feeling that it is not possible to insist on zero risk in the development of a new industry which has the potentials for contributing immensely to man's well being. Since there may be some degree of a biological effect associated with any exposure to ionizing radiation, this effect must be accepted as inevitable. But since we do not know that there can be complete recovery from the effects, it is necessary to fall back on an arbitrary decision about how much man will put up with — not a sceintific finding, but an administrative decision.

It should be emphasized that although the limit of acceptable risk of injury to people becomes the standard against which all factors are judged, it is not a "standard" which is approached as closely as possible (e.g., a standard of weight). It is a condition (e.g., a cliff) which one stays away from as far as conditions permit. The closer one approaches the edge the more careful he becomes. Against this framework the function of radiation protection criteria or standards is to guide the actions of people in relation to their environment. Their violation breaks the rule but does not define when one has been pushed over the cliff.

In discussing this general point, Parker stated in his summary — Standards are: (1) models for judgment as to whether a given action should be taken, (2) communications of the expert to guide those less experienced, (3) crystallizations of past experience utilized to facilitate future actions, and (4) means for integrating a whole made up of many parts.

He further pointed out two ways in which standards may originate and develop, especially in the radiation protection field. The first concerns moral and ethical standards, such as exemplified by the Hippocratic oath and the standard conduct of a physician. In application to radiation protection, concern for the well-being of our future generations in the face of the deleterious genetic effect of ionizing radiation must lead to standards of this kind. The second type of standard might be called an "action facilitator," a device to allow each individual in society to proceed in a certain manner, confident that certain obstacles have been removed. The rule of the road is a simple example. In just this sense, there is a class of radiation protection standards designed to permit an individual to proceed with reasonable safety for himself and with the assurance of not causing unreasonable interference with the needs, work, or safety of others.

There was substantial discussion about the use and meaning of the terms "threshold" and "linearity" in relation to the dose effects. Austin Brues, speaking sharply to these two terms, pointed out that the idea had been popularized that one or the other of these conditions has to exist while on the contrary it is essentially a foolish and unscientific procedure to choose the first two hypothetical guesses that come to mind and to set them off against each other. He indicated that unfortunately this is the way a great many people have been trained to think; they find it difficult to encompass more than one conflict of concept at a time. Even many scientists find it convenient to test one theory at a time and this is all right, provided it is used as a guide for further theory; but it is fallacious when it is done for the sake of decision making. (These observations have been noted because of the great furor raised nine years later by individuals who used the concepts as facts rather than hypotheses).

The hearings brought out clearly that the philosophy of risk, as indicated by the NCRP in 1949, was one that had to be lived with. With respect to economic and social concepts and the concepts of risks, there appeared to be general agreement among the witnesses on the following points:

1. The essentially ethical-social judgment that "no cost is too high to pay for safety," sometimes expressed by saying that cost is irrelevent to the protection of persons from ionizing radiation, provided (a) that the term "cost" related to direct expenditures for research in environmental monitoring shielding, etc. and (b) the judgment is qualified by recognition of the imperfect technology currently available but of the opportunities for protection in the future.[43]

2. In the more fundamental sense, the hypothesis that all man-made radiation carries some risk implies that (a) implicitly or explicitly some sort of balance of progress and risk is necessary and (b) a decision to proceed or not to proceed with the uses of atomic energy automatically implies such a balancing whether or not it is conscious.

3. It is essentially impossible to reduce to quantitative terms the dollar value of radiation protection.

4. It is even more difficult to quantitate the cost equivalent to any radiation exposure risk borne by persons in lieu of protection or to the benefits to be associated with any enterprise for which radiation protection may be required.

5. The NCRP has attempted to establish economic and social contributions in its formulation of basic radiation protection recommendations, mostly by asking the question, "Can such and such an industry 'live' with a proposed dose level?" There was some disagreement as to whether the NCRP recommendations applied only to peaceful purposes and not to the basic issues of national policy such as posed by fallout. It was clear, however, that entirely different kinds of social considerations would have to be considered for normal peacetime conditions as against other conditions where the governing factors might derive from questions of national defense.

6. The FRC had also attempted to include social and economic judgments in developing their radiation protection guides by submitting their proposals to the Federal agencies for their qualita-

tive judgment, but this also seemed to apply mainly to "normal peacetime operations."

7. Neither the NCRP nor FRC appeared to have given explicit consideration, for atomic energy uses covered by their recommendations or guides, to the types of decisions actually to be made, although the consensus was that some decisions would certainly be determined by the recommendations made.

There was general agreement on the following points with respect to applying the basic and derived NCRP recommendations and practical operations: (1) their application is mainly for normal peaceful uses and not for special military uses such as weapon testing and not for medical exposure to patients, (2) there seems to be no undue technological burden in keeping actual exposures as low as practicable and well within the NCRP occupational exposure and environmental recommendations, (3) exposures in excess of recommended limits cannot be considered necessarily harmful, (4) recommendations are under constant review by the NCRP which includes individuals able to speak for practical problems of applying them.

The NCRP recommendations while intended for use as decision-action standards are by no means self applying but require professional judgment for practical application.

Mr. Duncan Holaday pointed out in his testimony pertaining to uranium mining that the working standard for air concentrations for radon daughters has long been specified on a basis somewhat different than would follow directly from the NCRP recommendation. For technical reasons, direct application of the maximum permissible concentrations of radon and air could not be readily applied to the uranium mining problem and for that reason the American Standard was expressed in other terms. Nevertheless, much of the American Standard for uranium mines and concentrators is taken up with recommendations for interpreting the NCRP's values of maximum permissible concentration for the elements of interest (This question was dealt with in considerable depth at later hearings by the Joint Committee.)

Aside from the numerical similarity of the FRC Guides and NCRP Recommendations, some witnesses expressed the view that the FRC activities to date were essentially a duplication of the NCRP effort and especially so since many NCRP members had been called upon by the FRC in the course of their studies. The major Federal Agencies affected by the FRC memorandum have not been required to change their regulations or alter the numbers reflected in their radiation protection standards because they were already following the NCRP recommendations.

It seemed that whether or not the FRC had solved any major problems to date was a question of less concern to many witnesses than the basic question of what will happen to the NCRP now that the FRC has been established. What role will the NCRP perform in the future? Will there be any bar to membership and full participation on the NCRP by employees of the Federal Government, particularly those individuals long active in the NCRP who are now being called upon directly or indirectly to work in support of the FRC? After considerable discussion, it was generally agreed that the NCRP should not only continue to perform its traditional functions of scientific study and evaluation of technical data, but also that it should retain its independence unhampered by government, industry, or any other interest group. It was emphasized as being important that any question of conflict of interest with respect to active participation in NCRP affairs of scientists or government employees be resolved without delay.

For full details on the above hearings, reference should be made to three sources. (1) Selected Materials and Radiation Protection Criteria and Standards: Their Basis and Use;[158] (2) Radiation Protection Criteria and Standards: Their Basis and Use (May-June Hearing Record);[160] (3) Joint Committee Print, October 1960.[161]

The third series of hearings on radiation standards was held by the Joint Committee in June 1962 under the title "Radiation Standards, Including Fallout."[163] As the title implied, these were directed mainly at problems centering around fallout from nuclear weapons and the activities of the FRC in relation to their guides and action levels, i.e., levels of exposure for which corrective actions would be taken to one degree or another.

The principal problems that were brought up relating to radiation protection standards were as follows:

1. The major issues in the field of radiation protection standards still seemed to involve their application to the population from an increasing variety of sources, particularly those relating to radioactivity in the environment, for example, fallout, waste disposal, space application, etc.

Major deficiencies appeared to lie in population exposure standards.

2. Some fundamental questions related to exposure of the population appeared to be unanswered; for example, what purposes are the standards suppose to achieve? Just what are radiation protection standards going to do? Control sources? Control industrial practices? Eliminate a present threat to health? Prevent a future threat to health? If none of these, what? Under the philosophies of radiation protection, is there some fixed value of radiation exposure from all sources which cannot be exceeded without undue risk to health? As a matter of national policy or philosophy, should radiation protection standards be applied to all programs of the government including those required by National Security and new applications under development? Is it either necessary or desirable for population standards to have a fixed numerical relation to occupational standards?

3. Effective leadership for developing radiation protection criteria and standards to cover new operational problems, such as fallout, was not yet being adequately exercised by any government agency or group at the present or at the time. Assertion of initiative by the government in this area is essential to clarify present public confusion and to provide advance guidance for special situations which may arise in the future.

4. The fundamental concept that no radiation exposure should be accepted unless there are good reasons for doing so means that the radiation exposure standards must be related specifically to the purpose for which they were derived. The reasons for accepting exposure are related to many other factors including national requirements and social, ethical, and economic considerations and they should be stated and not hidden under the guise of health effects..

5. There continues to be no effective means by which social and economic factors can, in fact, be applied to the concert of other major considerations in the development of radiation standards, especially for application to fallout.

Reference should be made to the full hearing record, June 4, 1962 and the Joint Committee Print, September 1962.[164]

Another important set of hearings by the JCAE was on the "Radiation Exposure of Uranium Miners" held in May, June, July, and August 1967.[169, 170]

The special problems of radiation exposure of uranium miners had been known for many years but serious attempts to examine the standards by which the exposure would be controlled were not undertaken until the 1950's. During this period, attention was drawn sharply to the fact that many uranium miners were suffering from lung cancer and preliminary surveys indicated that the incidence was clearly higher than in the general population living in the same areas.

At the outset, the standards that had been adopted by the NCRP and ICRP for the control of radon were centered more about situations in factories, laboratories, processing plants, etc., and as already noted briefly above, could not be applied directly to the radioactive products that were present in the air of uranium mines. The difficulty centered in large measure around problems of measurement and interpretation. In 1959, the NCRP reduced its recommended level for radon by a factor of 3 and later in the same year the same value was recommended by the ICRP.[78, 108] Neither the old nor the new values was based on the special needs of a particular industry but on what was considered to be a maximum permissible dose rate to the critical body organ.

In the meantime, the Public Health Service working in conjunction with a special committee of the American Standards Association had proposed a new unit which it was felt could be applied on a practical basis. One unit was designated the "working level" (WL), defined as any combination of the short-lived radon daughters in one liter of air that will result in the ultimate emission of 1.3×10^5 MeV of potential alpha energy in their decay to lead-210 (Radium D). The exact conversion of working level to rads had contained some elements of uncertainty depending upon the models or assumptions used. One value suggested by the Federal Radiation Council would yield a dose of some 35 rads for minor exposed for 12 months to one working level.[142]

The original PHS publication and the FRC staff report made it clear that the working level refers to a combination of short-lived radionuclides polonium-218, lead-214, bismuth-214, polonium-214, also known as Radium A, Radium B, Radium C, and Radium C, respectively.[190]

Another unit that was introduced was the "working level month" (WLM). In discussing this, the FRC Staff Report stated that the inhalation of air containing a radon daughter concentration of 1

WL for 170 working hours results in an exposure of 1 WLM.

At the time that efforts were being made to reach an agreed acceptable working level time for uranium miners, there was substantial disagreement among members of the FRC, some choosing values as much as three times the other. Thus, the Federal Radiation Council recommended mine control, such that no individual would receive more than 6 WLM in any consecutive three-month period or exceed 12 WLM a year. At the same time, the Department of Labor stipulated a value of 1/3 WL per month as the upper limit.

At the time of the hearings, considerable discussion was centered about the possible existence of a threshold of effect for radium exposure. While there were no pressing claims made that there was a threshold, suggestive evidence was presented that there might be something termed a "practical threshold." This would be defined as a level below which no injury had been observed. This argument could follow from the analysis of a large number of individuals who had known amounts of radioactive material in the body and yet did not suffer any detectable injury. The reasonable opposing position was presented, that although no effect had been demonstrated at low levels of exposure, and effect could be present, but not demonstrated, because of the small population sample size. Without trying to decide the question of threshold one way or the other, there was general agreement that for a cumulative lifetime exposure below some 3 to 400 WLM's, the risk of incurring lung cancer form radiation was negligible. The question is not yet completely settled and is being studied intensively by various organizations.

For full discussions on the question of uranium miners, reference should be made to the Hearing Record and the Joint Committee Print of December 1967.[171]

CHANGES IN UNDERSTANDING OF GENETIC EFFECTS OF RADIATION

It has been noted above, in connection with the report by the National Academy of Sciences on the Biological Effects of Atomic Radiation, that beginning about 1958 experimental evidence began to develop which showed that the concepts on genetic effects expressed only a few years earlier were in serious doubt. This related particularly to the work of W. L. Russell. Continued studies by his group, as well as work in other laboratories, have verified the initial results and extended them further. The results were summarized in hearings held by the Joint Committee on Atomic Energy on the "Environmental Effects of Producing Electric Power."[172-174] The high points as presented by Russell will be outlined briefly.

The important question and issue were whether there might be a threshold dose or threshold dose rate of radiation below which radiation would produce no genetic damage at all. Prior to 1958, it was generally assumed on the basis of experimental work, particularly that on the fruit fly, that there was no such threshold for genetic effects. However, in his first paper in 1958, Russell showed that contrary to the report on the fruit fly, when a radiation dose given to mice is spread out in time and delivered at low dose rate, it produces less genetic damage than the same dose given in a short time high dose rate.[117,48] This finding has been reconfirmed repeatedly and has been checked independently at the British Medical Research Council's Laboratory at Harwell.

While there was considerable speculation among scientists regarding the validity of the results, it now appears that both sides of the controversy were right and both were wrong, depending upon which sex was being considered. As the radiation dose rate to which the female is exposed gets lower and lower, the mutation frequency produced by a given dose also goes down until, at the lowest dose rate tested, (a few R/wk) the mutation frequency is not significantly higher than that from the spontaneous mutation in the controls.

On the other hand, in the male, although to begin with the mutation frequency also drops as the dose rate is lowered, a point is reached at a dose rate in the neighborhood of one roentgen per minute, below which further reduction in dose rate has no effect. A given total dose, at dose rate as low as ten roentgens per week to the male, produces as much genetic damage as the same dose at a dose rate of one roentgen per minute (2400 R/wk). These compare with the average dose rate of 0.1 R/wk corresponding to the maximum of 5 R/yr allowed for radiation workers. Total doses in the order of 400 to 600 roentgens were delivered in each experiment.

It was further indicated that there can be both

somatic effects on the reproductive cell, for example, the killing of the cell itself or an effect on its division capacity, etc., or there can be a genetic effect on the reproductive cell which is passed on the the next generation. The mature spermatozoa in the male, for example, are highly resistant to somatic damage from radiation. It takes very high doses to kill them or even to destroy their fertilizing capacity. On the contrary, the early germ cells in the female are highly sensitive to killing.

Estimates of the genetic radiation damage are based on the cells that are more easily damaged genetically. This effect of the interval between irradiation and conception is a recent finding which indicates that in the human, there might be less hazard than was thought to be the case at the time when the permissible levels were established in 1956. Studies on the oocytes show some differences in response of the female in mouse and human to fertility effects. This raises caution about whether genetic effects in the female can be carried over with the same degree of confidence that had been established in the case of the male. Putting together the various pieces of information established over the 1960's, it has been concluded that the best estimate of the average genetic risk from exposure of both sexes at low dose rates, or low doses, is only about one sixth of what it was estimated to be in 1956.

Important deviations from the earlier concepts have also been established between groups of male mice irradiated as newborn. The mutation frequency in the offspring produced by these animals when they grew up to reproductive age appears to be somewhat below that obtained in the offspring of males first irradiated as adults.

Another entirely different type of experiment concerns the possible interaction between chemicals and radiation. The probable explanation for the dose-rate effect of radiation observed in the mice is that repair of mutational or premutational damage is going on at low dose rates. It has been found in other organisms that such repair processes can be damaged by certain chemicals, one of which is caffeine. Since this is a drug consumed in great quantities by man, it has been important to find out whether it had any effect on the mutational repair process in mice. Careful experiments have indicated that there is no effect, but the possibility remains for other chemical mutagens.

In spite of this relatively radical change in the results of genetic studies, with the indication that the risk at low doses and low dose rates is considerably below the early estimates, the magnitude and complexity of the effects were such as to not warrant serious consideration for raising the permissible dose levels. However, estimates of genetic damage based on earlier reports are in substantial error.

NCRP ACTIVITIES 1958 TO 1964

Note has already been made of the concern of the Director of the National Bureau of Standards over the relationships between the NBS and the NCRP. In 1957, a meeting of the full committee was held to discuss the future position of the committee. Dr. Astin attended the first part of the meeting to make known the reasons for his concern and to make some suggestions for future actions by the NCRP. In brief, his concern was that the NCRP did not fall into the pattern of government committee structures or government participation in committees as outlined in the Executive Order of about 1954. In support of this, he cited an opinion by the chief counsel for the Department of Commerce. While no immediate action by the Bureau seemed to be sharply indicated, he, nevertheless, felt that sooner or later the situation would become difficult and hence the NCRP should actively work out a solution. He offered two suggestions: (1) that the committee incorporate itself in some state or, (2) the committee seek a Congressional Charter.

To enable the committee to pursue the organizational problem vigorously, the committee decided to temporarily stop the normal rotation of membership on the Executive Committee and ask the current committee to continue until a solution had been reached. In about 1962, it was learned through informal channels that Congressman Chet Hollifield, Chairman of the Joint Committee on Atomic Energy was increasingly concerned over the possible disappearance of the NCRP as the only independent and nongovernmental protection body in the United States and that he might be willing to sponsor a bill in Congress to grant the NCRP a Congressional Charter. After some discussions this resulted in his introducing Bill HR 10437 to the 88th Congress in 1964. A companion Bill was introduced at the same time by Senator

Pastore. The matter was referred to the Committees on the Judiciary in the House and Senate.

The bills were reported out favorably and passed by the House on April 6, 1964, and by the Senate on July 2; on July 14, the Charter was established as an Act to "Incorporate the National Committee on Radiation Protection and Measurements," under Public Law 88-376 (78 Stat 320). The legislative history of these actions is contained in the House Report No. 1252 (Committee on the Judiciary) and Senate Report No. 1155 (Committee on the Judiciary), and in the Congressional Record, Volume 110 (1964).

The Charter of the Council stated its objectives as follows:

1. To collect, analyze, develop, and disseminate in the public interest, information and recommendations about (a) protection against radiation (referred to herein as "radiation protection"), and (b) radiation measurements, quantities, and units, particularly those concerned with radiation protection;

2. To provide a means by which organizations concerned with the scientific and related aspects of radiation protection and of radiation quantities, units, and measurements may cooperate for effective utilization of their combined resources, and to stimulate the work of such organizations;

3. To develop basic concepts about radiation quantities, units, and measurements, about the application of these concepts, and about radiation protection, and

4. To cooperate with the International Commission on Radiological Protection, the Federal Radiation Council, the International Commission on Radiological Units in Measurements, and other national and international organizations, governmental and private, concerned with radiation quantities, units, and measurements and with radiation protection.

The initial organization, now named the National Council on Radiation Protection and Measurements, had as charter members those who were members of the committee at the time of the enactment of the charter. Under the bylaws established in accordance with the requirements of the charter, a system of membership rotation was established with a normal term of membership by election to be for six years. At the outset, those members with names falling alphabetically in the first third of the total had membership of two years; those in the second third, four years; and in the third third, six years. Election of new members is affected by a quorum of the members present at the prescribed annual meeting. General management of the affairs of the Council rested in the Board of Directors consisting of ten persons, elected annually from among the members. The president would act as the Chairman of the Board and the vice-president would be a member ex-officio. The Board would be required to meet at least once a year, but in fact it has proven necessary to meet at least four times a year. By unwritten agreement, board members are reelected for total terms of about four or five years, with two rotating off each year.

In recognition of its responsibility to facilitate and stimulate cooperation among organizations concerned with the scientific and related aspects of radiation protection and management, the Council has created a category of NCRP "Collaborating Organization." Organizations or groups of organizations which are national or international in scope and which are concerned with scientific problems involving radiations, quantities, units, and measurements and radiation protection may be admitted to collaborating status by the Council. As of 1970 there were 28 NCRP collaborating organizations.

Programs and scientific committee activities in the new organization follows essentially the same structure and mode of operation that had been found effective for many years previously and outlined above. The need for any new activities may be brought to the attention of the Board by virtually any channel. This might be from suggestions from, (1) members of the Council or members of the various committees, (2) any of the collaborating organizations, (3) any other agency or organization, (4) through contract negotiation with some agency or organization, and (5) by anyone feeling that the NCRP could fill a particular need.

After due consideration of any suggestion or recommendation, the Board of Directors would make a formal decision whether or not to institute the program or study or establish a new committee.

As before, when a new committee is established, the chairman is selected by the Board either from among the council members or from outside the council. Members of the committee would be similarly appointed, all of these actions taking place after ballot or action by the Board of Directors. Upon completion of a report or study by a scientific committee, evidence to that effect is presented to the Board and if an initial review of the report indicates that it is adequate, it is then assigned to a number of "critical reviewers" for detailed examination on behalf of the Board. Depending upon their comments and recommendations, the report might then either be sent back to the committee for further work or be sent out to the entire Council membership for review and approval by ballot. While unanimous approval by all of the council members is not necessarily required, it is only rarely that a report is released for publication without unanimous approval of all of those casting ballots. It must be clear that occasionally strong differences of opinion will generate and in general it is the policy to either resolve these opinions or remove the points of disagreement from the report.

NCRP REPORT NO. 29

During the period while the NCRP was undergoing its reorganizational pangs, its committee activities continued without interruption and ten reports were completed and issued between the period 1958 to 1964; the first seven of these have already been mentioned above.

Report No. 29 on "Exposure to Radiation in an Emergency" was prepared under the Chairmanship of Dr. G. V. LeRoy and issued in January, 1962.[87] This report had as its main purpose the development of information as to the doses that might be accepted by several categories of civil defense workers under emergency conditions. It followed an interim report that had been forwarded in September 1954 and that was used by the Civil Defense Officials who always accompanied it by the statement that the information was of an interim nature.

In response to a formal request from the Director of The Federal Civil Defense Administration in 1955, the NCRP undertook a broad study of the emergency exposure problem with particular reference to conditions that might result from a nuclear attack. Initially, attention was focused on the extent to which the whole-body gamma radiation, (a) caused injury, (b) impaired the capacity to work, (c) reduced fertility, (d) caused late somatic effects, such as leukemia, life shortening, etc., and (e) caused genetic injury. The Committee examined, in detail, the problem of stipulating values for permissible doses for selected personnel engaged in tasks of varying priority during a post-attack period. It soon became apparent that this approach, however commendable in the case of a radiation accident in peacetime, was not realistic in a thermonuclear war. In retrospect it was evident that the magnitude of the situation was not fully appreciated when the Committee first examined the question of "the amount of radiation that might be accepted" (1955). Studies such as a Rand Corporation Civil Defense Report of 1957 convinced the subcommittee that the real problem was survival. The question then became "how much radiation can people stand?" — not how much is acceptable or permissible. The entire subcommittee shared with many others "the enormous psychological difficulty which everybody has in coming to grips with the concept of thermonuclear war as a disaster that may be experienced and recovered from."

After considerable debate within the subcommittee it was decided to take an entirely different approach and to prepare material directed primarily at what might be encountered in widespread civil defense operations. At the same time it was obvious that many of the same principles could be applied to less extensive civil radiation disasters under peace-time conditions. For such reasons as these the report was written at a level of understanding by the kinds of commanders and volunteers that might be found in a civil defense or disaster situation.

The problems of dealing with any situation as serious as nuclear attack involved many conditions with which the committee had not dealt before. Implicit in the circumstances of any emergency is the temporary loss of control over exposure to radiation so that some or all of the people involved are in danger of receiving doses in excess of "normal peacetime" permissible limits.

Any decision involving the controllable exposure of an individual to radiation under emergency conditions is intimately related to many factors involving human judgment under stress. Unfortunately, any decision involving radiation exposure is irrevocable once the exposure is received; thus, the officials in charge of emergency operations must examine any proposed action involving radiation exposure in relation to all other elements of the situation.

It is against this background that the process of decision making in a radiation emergency must be considered. Recognizing that the purpose of all radiation protection is to minimize injury, the committee made a series of recommendations. Some of these will be given below but without the supporting and pertinent discussion.

1. In peacetime emergencies, the objective of radiation protection is the fewest persons exposed and the least possible exposure to them, and

2. In war emergencies the objectives are first, the fewest deaths; second, the fewest requiring medical care; third, the smallest amount of genetic injury; and, fourth, the lowest probability of late somatic effects.

3. Make no allowance for recovery during the first four days but take the accumulated dose as equivalent to a single dose of equal size, and

4. Use the equivalent residual dose (ERD) for exposure protracted beyond four days. (The concept of ERD will be discussed below.)

5. The entire population must be considered equally susceptible to the effects of radiation.

6. No administrative distinctions should be made between injuries caused by radiation and other casualty-producing agents

7. The categories of outcome of exposure to radiation be limited to: (a) medical care not required, (b) medical care required during the emergency period or subsequently (except that late effects are not considered) and, (c) death.

8. All estimates of equivalent residual dose should be assigned the same percent uncertainty as the accumulated exposures on which they were based.

9. Any prediction of the number of casualties should be assigned an uncertainty of at least plus or minus 25%.

10. The possibility of genetic injury should not be a principal determining factor when making decisions during a war emergency.

11. The possibility of late somatic effects of radiation should not be a principal determining factor for making decisions during an emergency.

12. The equivalent residual dose should be used to plan protracted exposure.

In addition to the effects on human beings, there was a short discussion of the effects on livestock and agriculture, particularly in relationship to how these items might be used for food under emergency conditions.

Because in an emergency exposures may vary widely as to dose rate and total dose, as well as the time over which the doses are received, it is necessary to develop some mechanism for helping with this kind of situation in a pragmatic way. To accomplish this the principle of equivalent residual dose (ERD) was introduced.

ERD is a concept that was believed to permit a more reliable prediction of the biological and medical consequences of the exposure to radiation than is possible on the basis of accumulated dose alone. By definition, ERD is the accumulated dose corrected for such recovery as has occurred at a specific time. It is presumed that a person who has received a particular ERD will display approximately the same signs and symptoms of radiation injury as would be anticipated following a brief dose of the same size. The decision to use ERD to evaluate radiation exposure in an emergency was based on the following considerations:

1. It is not possible to predict the immediate effect of any amount of radiation unless one knows the manner and duration of exposure;

2. The body can repair a substantial fraction of the injury responsible for such immediate effects as acute radiation sickness;

3. Recovery requires time; and

4. What injury cannot be repaired persists and successive increments of the irreparable injury are cumulative.

Because quantitative information on the rate or extent of recovery in man is limited, it is necessary to make certain assumptions on the basis of experiments with animals. In order to evaluate the effects of large protracted exposures, such as may occur in emergency, the following assumptions were made:

1. Ten percent of the injury is irreparable and may cause late effects;

2. The body recovers from the reparable 90% of the injury in such a fashion that about half of the recovery has occurred in one month and nearly all possible recovery has occurred after three months;

3. The process of recovery is continuous in the case of protracted exposure;

4. Since there are no proper units to describe radiation injury, it is convenient to consider that a brief dose and the injury that it causes are proportional in magnitude.

The committee recommended that the ERD concept be used only to predict immediate effects such as radiation injury or acute radiation sickness of the kind expected following brief doses in the range below 300 R. Although it may seem practical to use the ERD to predict the additional protracted exposure that may be fatal, it was not recommended. Similarly there was no reason to believe that ERD was a reliable predictor of any of the late somatic effects of radiation or genetic effects. The principal advantage to be gained from using the ERD was to evaluate the combination of brief or protracted exposures that can be expected in most radiation emergencies.

The ERD when used within the restrictions specified was believed to be a pragmatic approach, useful in the decision-making process. However, some individuals have extended the use of the concept considerably beyond that envisioned by the Committee and under these conditions the end result could be misleading.

The Committee presented a formulation for making estimates of the equivalent residual dose and this carried the obvious appearance of an understanding and accuracy that was not justified. This resulted in varying levels of criticism, both from outside of the Committee and within the Committee itself. The subject was debated in depth during a series of meetings held in Interlaken in June 1969.[49]

The principal negative report on the concept was delivered by the chairman of the committee that had originated it and it was a considerable surprise to discover that many persons taking part in the discussions did not want to discard the principle unless something clearly better could be found. The original committee has since been reorganized and is currently considering the problem and trying to find an alternate method which will be technically more acceptable. It appears that while some modifications may be made, the ERD principle still fits the facts better than any other model.

NATIONAL ADVISORY COMMITTEE ON RADIATION (NACOR)

Another new radiation protection activity that was to have long range importance was the establishment in 1958 of the National Advisory Committee on Radiation. This was a strictly government type committee set up by the Surgeon General of the U.S. Public Health Service to provide him with guidance on matters pertaining to the control of radiation hazards in the United States. Among the many assignments given to the committee, was the task of evaluating the programs then underway in the United States to protect the health and well being of the public from the possible hazards of ionizing radiation. The committee was established under the Chairmanship of Dr. Russell H. Morgan and its 11 additional members, selected from the scientific community at large, were appointed for three year terms.

The first report of this committee was issued in March 1959 and dealt with (1) radiation hazard as a problem and public health,[50] (2) the elements of radiation control programs, (3) state vs. federal regulation of radiation protection, (4) radiation safety personnel, and (5) comments and recommendations. The following proposals and recommendations were submitted to the Surgeon General for review and appropriate action: (1) Primary responsibility for the nation's protection from radiation hazards should be established in a single agency of the Federal Government; logically this should be the U.S. Public Health Service. (2) The agency should be granted authority for broad planning in the field of Radiation Control including the coordination of state and local regulatory programs with the safety operations of federal and private groups in a manner which would provide a unified attack on problems associated with the control of radiation hazards. (3) The agency should be given the authority to develop a comprehensive program for *all* sources of radiation. In this connection the committee directed attention to the following principles and additional recommendations:

a. Radiation protection standards constitute a matter of broad national importance: problems of radiation control frequently do not respect state or regional boundaries, but extend across large areas of the nation, therefore, the committee recommended that the agency be charged with the responsibility of promulgating uniform national standards on radiation protection. To meet this responsibility the agency should take full advantage of the guidance provided by the National Committee on Radiation Protection (NCRP) and other organizations of similar character. Furthermore, the committee recommended that the agency be granted authority to undertake intensive research programs aimed directly at provision of scientific data for the development of improved standards of radiation protection.

b. Since the enforcement of regulations affecting the health and well being of our society has traditionally been the responsibility of state and local agencies, there appeared to be no fundamental reasons why these agencies should not bear substantial responsibility for the regulation of health hazards associated with radiation exposure. Therefore, the committee, recommended that as much regulatory responsibility as possible be vested within state and local governments in the field of radiation protection. However, so that the agency may be assured of discharging its responsibilities to the nation as a whole, it was recommended that it be granted supervening authority in those areas of enforcement where federal regulations seem more appropriate. Finally, so that state and local governments may discharge their responsibilities with the greatest effectiveness, it was recommended that the agency be granted authority to provide technical and financial assistance to such governments as in other public health programs.

c. The training of professional and technical personnel with which to meet federal, state, and local requirements over the years, is a problem of national importance, hence the committee recommended that the agency be granted authority to undertake a broad range of training programs which would assure that the national state and local needs for personnel trained in radiation protection, would be satisfactorily met.

It is also worth noting that the committee recommended one of the largest budgetary expenditures ever proposed to carry out such recommendations. This was suggested to reach a level of approximately $50 million in a period of five years.

A second report of the National Advisory Committee on Radiation was issued in May 1962, dealing especially with radioactive contamination of the environment and the necessary public health actions.[51] Its recommendations included the following:

1. The Public Health Service should substantially intensify its efforts to develop a comprehensive program to control environmental radiocontamination.

2. The Service should take immediate steps to expand its research and development programs in radiation control. In the field of surveillance, particular attention should be given to technical investigations to devise (a) improved methods of monitoring by which more rapid and detailed measurements could be made, (b) more reliable automatic surveillance equipment which could be operated with a minimum of highly technical manpower, and (c) standard techniques and methods which would yield consistently reliable results from laboratory to laboratory. Noting that in the field of countermeasures the research demands are particularly great, and since countermeasures to control environmental contamination were in a relatively early stage of development, the research effort should be directed to: (a) improve the safety of radiation sources, (b) control the vectors by which radio contaminants are distributed to the population, and (c) minimize or prevent the deleterious effects of radiocontamination to the body.

3. That training programs for radiation health specialists and radiological technicians should be fully supported and strengthened.

4. That the Service should be provided with funds to meet its broad responsibilities in radiological health.

The body of the report included discussions of the measurement of environmental levels of radiocontamination, evaluation of surveillance data, and countermeasures.

The third and last NACOR Report was issued in April 1966 and dealt mainly with the developing role of the Public Health Services in the uses and control of ionizing radiation and an evaluation of manpower shortages and related academic problems in the radiological sciences.[52] Shortly after the submission of its third report, general circumstances within the Department of Health, Education, and Welfare were such that the Committee could be discontinued.

While the activity of NACOR did not contribute directly to the development of new radiation protection standards or philosophy, it nevertheless played an extremely important role in alerting and activating the Public Health Service to its responsibilities in the field. It was to a considerable extent responsible for the development and buildup of what came to be designated as the Bureau of Radiological Health (BRH), with a widespread organization prepared to deal with many of the problems that might arise in the event of radioactive contamination of the environment. As will be noted later, the BRH has played an increasingly active role in the evaluation of potential radiation hazards in the country as, for example, the surveys of radiation exposure both in the open environment and in the medical applications of radiation. The Bureau was also to become a center for the inspection and evaluation of radiation-producing devices of all kinds.

It has already been noted above that the establishment of a viable radiation health program began in the U.S. Public Health Service in about 1956 under the initiative of Surgeon General Burney. Actually their concern with some radiation matters had begun as early as 1941 when they issued a report on "Radium and X-ray Hazards in Hospitals."[191] Other scattered studies were made up until September 1948 at which time a Radiological Health Branch was established in the Bureau of State Services. From this point on, the organization led a somewhat checkered growth pattern with a small staff carrying out scattered studies, mostly in the field, until the Branch became the Radiological Health Program in the Division of Sanitary Engineering Services. In 1956, in collaboration with the AEC, they established a National Radiation Surveillance Network.

Following the appointment of the National Advisory Committee on Radiation (NACOR) the activity became better organized and began to grow systematically and in 1959 the following responsibilities were assigned to the PHS: (1) research on the effects of radiation on living matter, (2) development of the means for collecting and interpreting data on all forms of radiation exposure throughout the U.S, (3) training of scientific professional and technical workers in radiological health, (4) technical assistance to Federal, State, and Local agencies, (5) development of recommendations for acceptable levels of radiation exposure from air, water, milk, medical procedures, and the general environment, and (6) public information and health education activities related to radiological Health.

In August 1960, they established a laboratory devoted principally to the development of methods for controlling radiation hazards from x-radiation. By this time the organization had become a Division of Radiological Health.

In January 1967, with a budget of some $20 million dollars, the Division became the National Center for Radiological Health and in October 1968, became the Bureau of Radiological Health as a component of the Environmental Control Administration, Consumer Protection and Environmental Health Service. Their total staff in Washington and the field was then some 800 persons and in July 1969, the Bureau was delegated the responsibility for the day-to-day administration of the Radiation Control for Health and Safety Act which had developed from the passage of P.L. 90-602. By the end of 1970, an entirely new Government Agency known as the Environmental Protection Agency (EPA) had been established and the Bureau was split into two parts, one remaining with the Public Health Service and the other going to the new EPA. Throughout most of the periods since the mid 1950's the radiation health programs had been caught in a series of regroupings and reorganizations within the Department of Health, Education, and Welfare but somehow survived and carried out important programs in the meantime.

ICRP ACTIVITIES 1958-1970

Note has already been made of the report prepared by the International Commission on Radiological Protection (ICRP) during the 1956 Meetings in Geneva. Immediately following these meetings, the Commission undertook the studies for UNSCEAR and thus began to schedule meetings more frequently than normal. As a result, addenda to the 1956 Report were made before it was finally issued in 1959. Amendements made during this period were published in several journals in 1957 and 1958.[104, 105] After further consideration, the ICRP recommendations were adopted in September 1958 and published in 1959 as ICRP Publication I.[107] Almost as soon as they were published, there were futher considerations given to some of the sections together with some new points of issue, and these were published for the first time in the preface to a report by Committee II on Permissible Dose From Internal Radiation that was published in mid-1959.[108]

These changes and addenda, made between the time the basic report was finished and the time that it was actually published, had, as noted, appeared in various ways with the result that the actual chronology is difficult to establish. Also during this particular period, the presentations were so confusing that in the United States there seemed to be a tendency to ignore the whole period. Fortunately, this situation was rectified about 1960 and has not recurred since.

In the meantime, both the ICRP and the ICRU entered into formal working relationships with the World Health Organization as a "Non-governmental Participating Organization." A similar relationship was established later with the International Atomic Energy Agency. In addition, there was close cooperation between the ICRP and other international groups such as the International Labor Office and Food and Agriculture Organization.

During its regular meeting in Munich in 1959, the Commission discussed further its basic recommendations contained in Report No. 1 and while no substantive changes were made, a number of explanatory statements were drafted.[108] Special attention was directed to the occupational exposure of pregnant women noting that they may present a special risk. They further indicated thay any special recommendations for pregnant women must, in practice, apply to all women of reproductive age. The Commission also noted that with the present maximum permissible exposure levels, no special treatment of radiation workers with respect to working hours and length of vacation, was warranted.

ICRP REPORT ON PERMISSIBLE DOSE FOR INTERNAL RADIATION

This report prepared under the Charimanship of Dr. K. Z. Morgan was, as noted above, essentially prepared in conjunction with the NCRP in the United States and has been discussed in connection with the NCRP reports.[108] Noting that the basic recommendations concerning radiation exposure have been revised in recent years by the ICRP, the Committee called attention to three changes of major interest to their treatment of the internal emitter problem.

1. Instead of a weekly maximum permissible dose, a quarterly limit was recommended.

2. While the permissible quarterly dose rates were essentially comparable to former permissible rates, a limit on integrated dose was imposed in the case of exposure of the blood forming organs and the gonads. The recommendations also apply the limit on integrated dose to the lenses of the eyes, but the relevant data were so inadequate that the eyes were not considered as an organ of reference in the Committee report.

3. Explicit recommendations were given for some non-occupational groups and limits were suggested for the whole population.

Because of these changes the new report reveals some extensive modifications and methods of estimating internal dose. Also the report contained about three times as many radionuclides as listed in the earlier publication. Refinements in the calculations for the exposure of the gastrointestinal tract and for chains of radionuclides in the body resulted in new values for many of the permissible limits. The power function model was discussed in the appendix as an alternative method for estimating the body burden for certain long-lived radionuclides. The data in the tables are expressed in terms of the exponential or compartment model for retention and elimination, but the maximum permissible concentrations (MPC) and body burden values listed in the tables were selected after careful consideration by the Committee of the values obtained by the use of both models.

All MPC values were given for a 40-hour week as well as for a continuous exposure of 168 hours per week. This was an added convenience because the values listed for continuous occupational exposure are convenient in obtaining permissible levels for special groups and for the population at large in accordance with recommendations in ICRP Report No. 1.

Although recognizing that the MPC values were, in many instances, based on very incomplete and, in some cases, uncertain data, they nevertheless were felt to embody the latest and best knowledge and constituted the best MPC values available. For many radionuclides, the radiation exposure period was recognized as possibly lasting for many months, or even a lifetime, although the intake might have occurred in a relatively short time. Thus, when radioactive contaminants were once deposited in the body, it was often difficult to make an accurate estimate of the total body burden or its distribution in the body. In most cases, even when the fact was established that a person carried a large internal burden of a radionuclide, little could be done to hasten its elimination from the body. However, in the light of present knowledge, occupational exposure for the working life of an individual at the maximum permissible value as recommended in the report is not expected to entail appreciable risk of damage to the individual or to present a hazard more severe than those commonly accepted in other industries. This is, of course, the general premise upon which occupational exposure hazards are based.

The 1956 decision of the ICRP to set the average external occupational exposure at 5 rems per year (corresponding to 0.1 rem per week), is not applied to internal dose calculations except in the cases of radionuclides that are distributed rather uniformly throughout the body or are concentrated in the gonads. The purposes of limiting the average weekly total body dose (0.1 rem) to one third of the former maximum weekly dose (0.3 rem) was to lessen the possible incidence of certain types of somatic damage, e.g., radiation induced leukemia and shortening of lifespan which were considered to result primarily from total body exposure. Obviously, the reduction in the gonad dose was intended to lower the incidence of deleterious genetic mutations that would give rise to effects appearing in future generations.

Inasmuch as the restriction of integrated dose applied primarily to the total body and gonad dose, there was no basic change in the permissible dose rate when individual organs such as the liver, spleen, bone, gastrointestinal tract, and kidney are the critical organs for the reasons given in the 1956 report. This was primarily based on the fact that when exposure is essentially limited to one organ, because of more-or-less selective accumulation of a certain radioactive isotope therein, it is obvious that this limit embodies a lesser risk when the whole body is exposed at the same permissible limit. For this reason, and the fact that none of the organs was known to be as sensitive as blood forming organs, gonads or lenses of the eye, the Commission decided to retain the previously recommended maximum permissible dose of 0.3 rem per week for each organ singularly.

It was noted, as in previous reports, that because the direct estimation of the body burden or of the dose to an organ or to the total body was generally difficult, and because in most cases measures to decrease the body burdens were rather ineffective and difficult to apply, the only practical procedure for general protection of occupational workers was to limit the concentration of the various radionuclides in the water, food, or air available for consumption. Examples were given for conditions involving combinations of internal and external exposure delivered over various times. During this period, the ICRP was adhering to its earlier recommendation giving quarterly permissible dose limits.

As already noted the ICRP during this period had a rather more elaborate breakdown of individuals according to how they might be exposed, either in occupational or nonoccupational situations or in borderline conditions. The treatment of these various conditions led to additional complexity in the internal dose problem.

In addition to the general discussion and explanation of the calculations, three fourths of the report was devoted to various tables listing permissible body burdens, permissible concentrations, and much of the tabulated information needed for the calculations. An extensive bibliography was included.[108]

ICRP REPORT ON "PROTECTION AGAINST X-RAYS UP TO ENERGIES OF 3 MeV AND BETA AND GAMMA RAYS FROM SEALED SOURCES."

The report of the Committee on "Protection against X-rays up to Energies of 3 MeV and Beta and Gamma Rays from Sealed Sources" was prepared under the Chairmanship of Dr. R. G. Jaeger and its report was published in 1960.[109] In general it should be regarded as a detailed updating of the earlier report issued in 1955.[104]

The report, based upon the codes of practice in operation in various countries, put special emphasis on the basic protection requirements leaving the expansion of the technical requirements to national committees. It did, however, add an extensive appendix containing graphs, tables, and examples from which the necessary numerical values for radiation protection could be obtained. In reference to the previous report,[104] new sections were added on the use of sealed beta and gamma ray sources as well as gamma-ray beam equipment. In addition, a subsection on x-ray analysis equipment was considerably expanded.

The general recommendations also recognized that with the increasing uses of radiation and radiation producing equipment, it was essential to consider not only radiation workers and patients but also all other persons in the vicinity, thus requiring lower permissible levels for those special groups. Noting the genetic consequences of the medical uses of ionizing radiations, special attention was given to the importance of avoiding unnecessary exposure of patients and recommendations were given to this end.

It was emphasized that the radiological examination of human beings for non-medical purposes as undesirable and not recommended. It gave examples of such examinations as shoe fitting and anti-crime fluoroscopy. It also gave further consideration to the emission of x-rays from television equipment and a reduction of the dose rate close to the surface of the television tubes was recommended.

In addition to the general medical applications, a series of recommendations was given for miscellaneous non-medical uses of radiation such, as already mentioned, radiation resulting from the irradiation of materials in x-ray microscopy, crystallography, etc., shoe fitting fluorscopes, and non-medical exposure of human beings. In addition, radiation emitted as an unwanted byproduct was recognized and some general requirements given for dealing with this category of radiation sources. This includes items such as electron microscopes, cathode-ray tubes, rectifiers, television and image tubes, etc.

Additionally recognizing that sealed beta and gamma ray sources are used extensively in industry, it was made certain that the recommendations were consistent with those for medical applications. Extensive appendices included curves and tables for the calculations of protective barriers, leakage radiation, scattered radiation, distance protection, protection against electrons, and the estimation of patient dose.

ICRP REPORT ON "PROTECTION AGAINST ELECTROMAGNETIC RADIATION ABOVE 3 MeV AND ELECTRONS, NEUTRONS, AND PROTONS."

This report was adopted in 1962 but had additions made to it in 1963, and was finally published in 1964.[1,1,2] It started out under the Chairmanship of Dr. W. V. Mayneord and a draft was completed about 1956, at which time the Committee came under the Chairmanship of Professor H. E. Johns. Because work was still needed, it was necessary to undergo a final editing by Dr. B. M. Wheatley.

Recognizing that although the hazards associated with the use of high energy x-rays and heavy particles, including neutrons and protons, are in many ways analogous to those occurring with x-rays produced at lower voltages, there are special risks and circumstances which call for detailed separate discussions. For practical reasons, 3 MeV was chosen as a dividing line between the interest of the Low Voltage X-ray Committee and the High Voltage X-ray Committee. In addition, the report extended to neutrons and protons with energies up to 1000 MeV.

While the current report was in preparation, the ICRU in its report No. 10A had made recommendations regarding the relative biological effectiveness of different radiation.[53] In its 1959 report, it had expressed misgivings over the utilization of the same term "RBE" in both radiobiology and radioprotection. Therefore the ICRU recommended that the term "RBE" be used in radiobiology only and that another name be used for the factor dependent on linear energy transfer for radiation protection. Needed was a quantity that expressed on a common scale, for all ionizing radiations, the irradiation incurred by exposed persons; the name recommended by the ICRU for this factor was the "Quality Factor (QF)."[53] Provisions for other factors were also made. Thus, a distribution factor (DF) may be used to express the modification of biological effect due to nonuniform distribution of internally deposited isotopes. The product of absorbed dose and modifying factors is termed the dose equivalent (DE). As a result of discussion between the ICRU and the ICRP, the following formulation was agreed upon.

1. For protection purposes it is useful to define a quantity which will be termed the "dose equivalent (DE)."

2. DE is defined as the product of absorbed dose, D, quality factor (QF), dose distribution factor (DF), and other modifying factors. (DE) equals D(QF) (DF)...

3. The unit of dose equivalent is the "rem." The dose equivalent is numerically equal to the

dose in rads multiplied by the appropriate modifying factors.

The report gave a table for the LET-QF relationship starting with a QF value for X and gamma rays as unity and stating that for electrons, it is only greater than unity at very low energies. (This latter statement was rescinded in 1970.)

For practical applications, a section on permissible flux densities was given for neutrons and protons, excluding thermal and intermediate energy neutrons.

The sections discussing the principles regarding working conditions and monitoring, and recommendations on equipment and operating conditions, followed generally along the line indicated in the basic report.

ICRP REPORT ON "HANDLING AND DISPOSAL OF RADIOACTIVE MATERIALS IN HOSPITALS AND MEDICAL RESEARCH ESTABLISHMENTS."

The report on the handling and disposal of radioactive material in hospitals and medical research establishments was prepared under the direction of Dr. C. P. Straub and was issued in 1965 after final editing by W. Binks.[113] It covered three categories of exposed individuals: 1. Individuals who are occupationally exposed to radiation, 2. Adults who work in the vicinity of controlled areas or who enter controlled areas occasionally in the course of their duties, but who are not themselves employed in work involving exposure to radiation, and 3. Individual members of the population at large, including persons living in the neighborhood of controlled areas. This was a change from earlier recommendations.

The report dealt with three principal aspects of handling and disposing of radioactive materials as they relate to hospitals in small laboratories: 1. the facilities to be provided in isotope laboratories and clinic and in wards and operating theaters, 2. the procedures to be adopted in handling, storing and transporting radioactive materials and, 3. the procedures to be adopted in the safe disposal of radioactive waste products and contaminated materials.

The report was generally a substantial upgrading of the material prepared in the 1950 reports. Responsibility for radiation safety measures was recommended as being in the hands of the authority having administrative control of an establishment; pre-employment medical examinations, and examinations during employment ere recommended, the latter depending upon the conditions of work. Environmental monitoring was recommended before operations with radioactive materials commenced, and then afterwards from time to time as conditions indicate.

Radionuclides were classified into four groups according to the relative radiotoxicity per unit of activity. This was based on the classification drawn up by the IAEA. Similarly, laboratories were grouped into four categories depending upon the level of activity and the type of material handled. Multiplying factors were given for a range of operating conditions. General recommendations were given for the kind of facilities needed for work with unsealed radioactivities including working procedures with such materials, their contamination control and actions to be taken in the case of spills, whether of a minor nature or of an emergency nature, such as in a fire.

Recommendations were also given on decontamination of personnel, buildings, equipment, and clothing, and for the disposal of liquid, solid or airborne wastes. Recommendations for the discharge of hospital patients during treatment with radioactive materials was left to national authorities, depending upon their local requirements. Some general suggestions were presented for the disposal of radioactive corpses.

A series of six appendices covered, 1. methods of testing sealed sources for leakage or surface contamination, 2. sterilization of small sealed sources, 3. description of an intermediate level laboratory facility, 4. radioactive contamination of surfaces, 5. example of a procedure for dealing with a serious spill, and 6. management of radioactive wastes.

"RECOMMENDATIONS OF THE ICRP AS AMENDED IN 1959 AND REVISED IN 1962."

The ICRP report entitled "Recommendations of the International Commission on Radiological Protection" (ICRP Publication 6).[114] was essentially a consolidation of Report No. 1,[107] together with the various amendments adopted since. Most of the points in this report have been covered in discussion of the earlier reports and amendments.

In its discussion of "dose rate effects" the Commission indicated that it appears possible, on some theoretical and experimental grounds, that when either the total dose or dose rate is very low, any effects will be directly proportional to the total dose and independent of the dose rate. This assumption had been explicit in past recommendations on permissible levels and although confirmatory proof is lacking, it is still believed to be a reasonable basis for assessment. For genetic effects, the report pointed out that linear dose-effect relationship unaffected by dose rate had been generally accepted in the past for gene mutations. However, they recognized that while the evidence for this was no longer as firm as a few years previously, the Commission did not plan at the time to modify its recommendations to allow for dose-rate effects in man.

There were some modification of the categories of exposure as outlined in their 1958 recommendations. It will be recalled that the categories of occupationally exposed individuals were divided into three groups of which one group was subdivided into three subgroups, In the current report, they simplified the subdivision as follows: 1. individuals who are occupationally exposed to radiation, 2. adults who work in the vicinity of controlled areas, or who enter controlled areas occasionally in the course of their duties but who are not themselves employed in work involving exposure to radiation, and 3. individual members of the population at large including persons living in the neighborhood of controlled areas.

Special attention was given to the exposure of women of reproductive age. For occupational exposure, the current report recommends that when a pregnancy has been diagnosed arrangements be made to insure that the exposure of the woman be such that the average dose to her fetus during the remaining period of the pregnancy does not exceed 1 rem.

For the radiological examinations of women of reproductive age, the report recommends that all radiological examinations of the lower abdomen and pelvis of women of reproductive age not of importance in connection with the immediate illness of the patient be limited in time to the period when pregnancy is improbable. The examinations that should be delayed to await the onset of the next menstruation are those that could, without detriment, be delayed until the conclusion of a pregnancy or at least until its latter half.

In its 1958 recommendations, the Commission had referred to an upper limit of the estimates of the annual genetically significant dose from medical exposure and noted that the highest levels could be reduced significantly by careful attention to techniques. While it is still clear that the allowed genetic dose may be related to the value of the medical benefit to be derived, future extensions in the use of radiological methods may confer greater benefit from their application than detriment from the necessary associated exposure of radiation. At the present time, therefore, the Commission decided to maintain its policy of not making numerical recommendations with regard to any appropriate genetic dose from medical exposure.

In consideration of the problem of emergency exposure of a local population after an accidental release of large quantities of radioactive material, the Commission noted that there is little experience upon which to base formal recommendations. It reaffirmed its former reliance on the 1959 report of the British Medical Research Council and extended similar recognition to the 1961 report as being a useful evaluation of the problem on the basis of the current limited knowledge.

The Commission reiterated its desire not to make any firm recommendations as to apportionment of the genetic dose to the population but nevertheless gave an illustrative case based on a 5 rem annual exposure.

In addition to modifications of the basic 1958 report, the current report provided some supplementary recommendations related to the report No. 2 on "Internal Emmiters" and Report No. 3 on "Protection against X-rays Up to Energies of 3 MeV."

ICRP REPORT ON PRINCIPLES OF ENVIRONMENTAL MONITORING

A report on the "Principles of Environmental Monitoring" (ICRP Publication No. 7) related to the handling of radioactive materials was prepared under the Chairmanship of H. J. Dunster and issued in 1966.[115] This report was more of an administrative guide to the establishment of protective methods and procedures as they might apply to environmental monitoring programs. It also introduced the important concept of the "critical group." Basically, it covered 1. the question of surveys outside of installations, the handling of radioactive and preoperational surveys, 2. emergency surveys, and 3. surveys for fallout of debris from nuclear explosions. Characteristic factors affecting each of these categories were outlined.

REPORT MADE TO THE ICRP

In the early 1960's, the ICRP decided that there would be advantages in having two kinds of reports included in its publication series. The first of these would be reports such as the first seven already noted and which are essentially recommendations, studied in depth by the Commission and issued for general use or adoption as desired. For identification they had traditionally appeared in brown covers. The second kinds of reports were to be prepared by special task groups dealing with subjects of interest and concern to the Commission but were published primarily for informational purposes and to provide summaries and solicit comments from the public. They were not necessarily reviewed by the full Commision in depth and were not intended by the Commission as official pronouncements. To distinguish them from the formal recommendations of the Commission, they bore a prefatory statement explaining their purpose and to be further distinguished appeared traditionally in blue covers. Unfortunately, the clear distinction between the nature of the two reports has been frequently misunderstood by some of the scientific public but the Commission should not be faulted for this misunderstanding.

ICRP PUBLICATION NO. 8: "THE EVALUATION OF RISKS FROM RADIATION

A report to the Commission entitled "The Evaluation of Risks from Radiation" was prepared by a Task Group under the Chairmanship of Dr. R. Scott Russell and was published in 1966.[116] It was first of the "blue" reports. The principal endeavor of this report was to present the available information on radiation effects as a function of dose with special reference to the low dose region; it dealt with both somatic and genetic risks.

Under somatic risks, the report dealt with information on the incidence of cancer as a function of dose with special attention being given to leukemia, thyroid carcinoma, bone sarcoma, and other neoplasms. Similar attention was also given to radiation induced developmental abnormalities, nonspecific reduction in lifespan, and other effects.

Under the genetic risks the report listed the genetic components of harmful variation in man, spontaneous and induced mutation detriment in man and mouse, and direct evidence from man of effects on offspring following irradiation of parents. Further sections dealt with induced changes in the somatic chromosomes of man and the problem of expressing detriment in terms which are meaningful. The big problem, of course, was to try to find some basis for evaluating the genetic risks to man, so the report discussed the bases for prediction of such effects and outline the dose mutation relationships which might be used as a basis of prediction or evaluation of risks in man.

In presenting its conclusions, the Task Group pointed out that it is more reasonable to suggest only the range within which effects may lie and that this can conveniently be done by defining "orders of risk." For example, a fifth order risk implied that the probability of an event (e.g., death or injury) to any individual is in the range of 1×10^{-5} to 10×10^{-5}, that is to say 10 to 100 injuries would be expected per million exposed persons. It was pointed out, however, that the emphasis which they gave to the limitation of the current assessment should not be regarded as implying that knowledge of the effects of low levels or radiation is any less precise than the effects of other toxic agents or environmental factors which cause similarly infrequent effects.

The data which were available on the effects of radiation mainly relate to the exposures of high dose-rate received during a limited period, and all estimates of effects of low doses and low dose rates are made on the basis of a linear dose-effect relationship below the doses at which quantitative information has been obtained. They pointed out that it must be borne in mind that in some instances this might lead to a gross overestimate of the incidence of effects from chronic low-level exposure and indeed that some of the effects may not occur at all. This made it apparent that the report could provide no simple solution to the practical dilemma in setting precise criteria for radiation protection.

In the area of somatic effects, studies of the late effects of radiation on human population show that when a substantial part of the body is exposed, the increased incidence of acute leukemia and chronic myeloid leukemia is the major fatal consequence. But at the same time there was no clear evidence from the human studies that radiation in small doses will give rise to appreciable "non-specific life shortening" or to serious consequences to the exposed individual other than cancer. For cancer, estimates were given of (sic) "the annual risk under continuous exposure to 1 rad per year or to the lifetime risk from a single exposure to 1 rad." This was described as a fifth order risk.[116]

With regard to the genetic risks of irradiation, it was recognized that the risk from small doses is much more difficult to assess than somatic risks. However, estimates were made of the detriment to the first-generation offspring of exposed persons, and on the basis of the linear dose/mutation relationship for gene mutations, it was estimated that the risk was of the third or fourth order. In an attempt to make a comparison of risks, it was felt implicit from the earlier discussions that no reasonable comparison can be made between the genetic burden imposed by radiation and that to which the population is currently subject as the result of industrial and social causes.

The report is an excellent summary of the information then available and is much too detailed to discuss in any more depth. Its references are extensive.

RECOMMENDATIONS ON RADIOLOGICAL PROTECTION(1965)
ICRP PUBLICATION No. 9

This report was prepared by the full Commission and was released in 1966.[117] It is a complete revision and updating of the 1958 report and while it carries over many of the earlier features, it also introduces some new ones and extends the results very substantially. The first section discusses the basic principles underlying the recommendations of the ICRP, giving the broad basis for the objectives of radiation protection, and a general discussion of the somatic and genetic effects against which protection is being sought.

A new section discusses the introduction of the concept of "dose equivalent,"[53] taking into consideration quality factors which are devised for protection purposes, as against RBE for which the application is limited to experimental purposes.

For the first time the Commission presented a statement on the broad concept of risk. While this is by no means new to the Commission's philosophy it is presented more comprehensively than in any of the previous official recommendations.

The report defines several categories of exposure constituting a considerable simplification of the former categories, for example, adults who may be exposed in the course of their work, or members of the public. For the latter, the term "dose limit" is applied (rather than Maximum Permissible Dose) to describe the radiation levels within which radiation should be controlled. For occupational exposure the same permissible doses previously recommended were continued. Special recommendations were given for persons whose previous exposure was unknown or for persons exposed in accordance with the old "maximum permissible doses," persons starting work at an age of less than 18 years, exposure of women of reproductive capacity, and exposure of pregnant women.

A new section was introduced dealing with exposure of the population including the assessment of genetic dose, the genetic dose limit and contributors to the genetic dose. The Commission noted that the illustrative example of apportionment which was given in the 1958 report may have come to be regarded as carrying a greater significance than was originally intended, but withdrew it in this report. In its place they discussed the genetic dose considerations as above and, in addition, dealt with the other principal sources: occupational exposure, medical exposure, and miscellaneous exposures.

Another new section dealt with action levels for exposures from uncontrolled sources, including abnormal exposures of radiation workers and abnormal exposure of population groups. The recommendations necessarily had to be schematic and again they referred to the Medical Research Council's report, "Hazards to Man of Nuclear and Allied Radiation."[54]

In earlier reports the ICRP had defined or described a maximum permissible dose; the current report (Pub. No. 9) was notable for its omission of this definition or its replacement by a new one. The term "dose limit" was specifically left undefined by action of the Commission when it adopted the term in 1965. At that time, it was made clear that maximum permissible dose (MPD) would apply to radiation workers and dose limits would apply to others, but the ICRP has slipped into the use of the term "dose limit" for all kinds of radiation exposure conditions.

ICRP PUBLICATION NO. 10: EVALUATION OF RADIATION DOSES TO BODY TISSUES FROM INTERNAL CONTAMINATION DUE TO OCCUPATIONAL EXPOSURE

The ICRP recommendation on the "Evaluation of Radiation Doses to Body Tissues from Internal Contamination Due to Occupational Exposure," a "brown" report, was prepared under the chairmanship of Dr. G. C. Butler and published in 1968.[118] The task group was charged with the amplification of paragraph 86 of ICRP Publication No. 6:

"Tests should be performed to estimate the total body burden for workers who deal with unsealed or active isotopes that may give rise to levels of ingestion, or inhalation, in excess of the maximum permissible concentrations. Such tests should also be performed where radioisotopes may enter the body through the skin or through skin punctures and open wounds. These tests may require the monitoring of breath and excreta and the determination by means of a total body monitor according to the circumstances. The radiation doses delivered to the appropriate organs or tissues should be calculated and noted on the personnel record and the permitted doses of external radiation should be adjusted to allow for the internal doses."

The group dealt with this very complicated problem by first discussing the general question of metabolism of radionuclides and their modes of entry into the body as, for example, by inhalation, ingestion, absorption through the skin, or entry through a wound. Necessary, also, was the development of an understanding of the translocation and disposition of radioactive materials, mainly n the vehicle of extracellular fluid by which transportable materials are transferred from one part of the body to another.

The bulk of the report was devoted to specific examples and calculations for a series of more commonly used radionuclides. The discussion of each nuclide is well referenced. The report also introduced the new concept of "Investigational Level."

ICRP PUBLICATION NO. 11: RADIOSENSITIVITY OF TISSUES IN THE BONE

A report giving a review of the "Radiosensitivity of the Tissues in Bone" was prepared by a task group under the chairmanship of Dr. J. F. Loutit and published in 1968.[119] This was a "blue" publication and hence did not contain specific Commission recommendations,

The terms of reference of the Task Group had been defined as: (1) the location and relative numbers of radiosensitive cells and cells not considered to be radiosensitive in bone and bone marrow; (2) the localization of bone-seeking radionuclides and of the ionization patterns resulting from their deposition in relation to radiosensitive cells of bone and marrow, (3) changes in these patterns resulting from growth and remodeling of bone, (4) biological data, especially on tumor formation and degenerative changes relevant to the above, and (5) comparative consideration of tumorigenic effects of external and internal irradiation of bone.

The Task Group dealt with the first four items in the context given to them in their terms of reference.

The fifth section on comparative tumorigenic

effects attempted to evaluate what human experience was available on effects of internal and external irradiation of bone. This included the ABCC studies of the Hiroshima and Nagasaki bomb survivors, the studies of American radiologists, British radiologists, the British cases of ankylosing spondylitis, and children irradiated for enlarged thymus. The main experience with internal irradiation was that of the radium dial painters. Studies of the experimental results on animals included osteogenic sarcoma and leukemia. The report discarded the old concept of bone being simply a piece of single tissue and pointed out that it should be considered as a complex of separate tissues for which individual dose limits would apply.

The summary and conclusion included the following: (1) Carcinogenesis (including leukemiogenesis) is the dose-limiting factor for bone. (2) Tissues considered at risk are osteogenic tissue - particularly on endosteal surfaces, haematopoietic marrow, and epithelium on bone surfaces. (3) Bone seeking radionuclides may be considered as far as distribution is concerned in four groups. (4) Under certain circumstances plutonium and thorium have a significant marrow distribution. (5) Osteogenic cells on endosteal surfaces can receive a high radiation dose, owing to the pattern of retention of plutonium and thorium in the marrow. Bothe haematopoietic marrow and osteogenic cells are at risk. (6) In man it appears that the tissue mainly at risk from external radiation is the haematopoietic marrow, whereas from radium it is the osteogenic tissue and epithelium adhering to bone. (7) In view of what is now known about the tissues at risk in association with bone it seems logical to consider a fresh approach to the calculation of maximum permissible levels in man. Such an approach now seems feasible for beta emmiters.

ICRP PUBLICATION NO. 12: ON GENERAL PRINCIPLES OF MONITORING FOR RADIATION PROTECTION OF WORKERS

The ICRP recommendations on "General Principles of Monitoring for Radiation Protection of Workers" was prepared by a Task Group under the chairmanship of H. J. Dunster and was published in 1969.[120] Its objective was to outline the principles of monitoring for radiation protection of workers including the monitoring of work places, personnel monitoring, and the relationships between the two. A special reference was made to specific paragraphs in ICRP Publication No. 9.

The report was primarily interpretive in nature, giving emphasis to the administrative aspects of personnel monitoring.

ICRP PUBLICATION No. 13: RADIATION PROTECTION IN SCHOOLS FOR PUPILS UP TO THE AGE OF 18 YEARS

The ICRP report on "Radiation Protection in Schools for Pupils up to the Age of 18 Years" was prepared by a Task Group consisting of E. D. Trout and E. E. Smith, for submission to Committee 3 of the ICRP under the chairmanship of Dr. B. Lindell.[121] The Commission distinguishes between adults exposed in the course of their work and individual members of the public for whom the annual dose limits are 1/10 of those for radiation workers.

In approaching the problem the report points out that "in many countries of the world, science courses in schools for pupils up to the age of about 18 years are tending to include, to an increasing extent, demonstrations and experiments involving sources of ionizing radiations. As large numbers — perhaps the majority — of pupils are likely to be affected, their exposure could contribute significantly to the population dose. In fact, a situation is likely to develop akin to that of medical radiology because in the fullness of time virtually everyone in the population will have received some exposure from this source. This is particularly important because the exposure occurs at an early age, which means that the gonad irradiation makes a maximum contribution to the population genetic dose."

The objective of this report as regards "school

exposure" is to achieve a situation whereby annual doses received by individual pupils in the course of instruction will be most unlikely to exceed 1/10 of the dose limits recommended for individual members of the public. The annual dose limits recommended for "school exposure" were therefore set as follows: for the gonads and red bone marrow — 50 mrems; for skin, bone, and thyroid — 300 mrems; for hands and forearms, feet and ankles — 750 mrems; and other single organs — 150 mrems.

(It will be noted that the basic permissible doses for critical organs is only one half of that specified previously by the NCRP).

Administrative aspects of this radiation control problem in school was outlined and a number of specific recommendations were made for the use of typical radiation-producing devices used in the course of teaching or school research.

ICRP PUBLICATION NO. 14: "RADIOSENSITIVITY AND SPATIAL DISTRIBUTION OF DOSE"

The ICRP report on "Radiosensitivity and Spatial Distribution of Dose" was prepared by two Task Groups, one under the chairmanship of Dr. L. F. Lamerton, and the other Dr. R. H. Mole. The report was published in 1969.[122] This is one of the reports to the Commission (blue cover) hence does not contain specific recommendations. It is an extremely valuable one in terms of the information that has been gathered, its evaluation and cross comparisons, and the essential references.

A summary of the conclusions on the spatial distribution questions includes the following:

1. The limitation of the present critical organs in concept, from the point of view of formulating rules for non-uniform exposure, was discussed and the possibility examined of modifying the system on the basis of risk estimates. It was concluded that no new system can reasonably be put forward at this time for various reasons including lack of knowledge of the relative radiation susceptibility of various parts of the body. It was suggested, however, that the exercise of developing a new scheme should be kept under review and as more information comes to hand, and new concepts in radiation protection are developed, the new approaches should be tested. One point emerging from the analysis is the need for decision on the basic standard for radiation protection — whether it refers to the whole body uniformly irradiated or to bone marrow.

2. The assumption of a linear dose-effect relationship and the range of dose and dose-rate over which it might reasonably be expected to hold were examined and the different classes of non-uniform exposure were considered in relationship to the use of mean tissue dose and to the concept of significant area. This suggested that: a) for partial irradiation of a tissue, the part irradiated is representative of the whole organ or tissue; a dose averaging can be carried out for local doses up to 100 rems in a year and possible higher. This applies to both skin and bone marrow. b) A "significant area" for skin of 1 cm^2 applicable to a small area would seem to be reasonable on grounds of operational convenience, and it is suggested that the present limit of 30 rems in a year to 1 cm^2 of skin be increased to at least 100 rems in a year.

3. Exposure from radioactive material, in particulate form, was discussed with special reference to the problem of plutonium in the pulmonary lymph nodes following inhalation of plutonium particulates. At present there appears to be no refinement of dose calculations which will deal with the problem and it was suggested that until direct experimental data are available dose limits should be derived on the normal basis by calculation of the mean dose throughout the whole organ or tissue mass, which in this case is all the lymphoid tissue.

4. Additivity procedures were considered, and the desirability of flexibility in the recommendations was again emphasized.

5. The general conclusion of the Task Group was that there is no case for attempting to set up highly elaborate rules for spatial non-uniformity of exposure. It is important to recognize that the dose limits are put forward as largely arbitrary norms of good practice, and there is virtue in recommendations which are administratively simple and have a degree of flexibility.

The report on Radiosensitivity of Different Tissue included some of the following general conclusions: (1) The kinds of radiosomatic damage which have to be considered when setting dose limits for occupational exposure at the present time are: cataract, impaired fertility, tumor induction, and defective development of the fetus. The possibility of impaired functioning of organs and tissues other than the gonads is so uncertain that it can be disregarded. The evidence for life shortening from effects other than tumor induction is inconclusive and not usable quantitatively. Sufficient is known now about irradiated human populations and the delayed effects which have occurred for it to be very unlikely that a significant hazard from exposure at current dose limits have been overlooked. (2) The dose response relation for each different kind of radiation effect should be considered on its own merits. There is good human evidence that there is a steeply sigmoid relation for opacification of the lens and that there is a threshold exposure below which opacities of a degree to interfere with vision do not occur. There are good general grounds for believing that the dose response relation for impairment of fertility is also sigmoid, but with a long tail due to a small fraction of persons of each sex who are much more sensitive than the average. Such scanty evidence as there is agrees with the working hypothesis that cancer induction in man may be taken as linear with dose. (3) The human evidence suggests that the annual dose limit for exposure of the lens of the eye should be 15 rems for both low and high LET radiation. (4) There is no human evidence on the response of the gonads to protracted irradiation which could be used to confirm the recommended dose limits and there is no reason for suggesting a change in them. (5) There is no general hypothesis about carcinogenesis which would allow reasonable predictions of tissue sensitivity to tumor induction by irradiation from known properties of tissues. (6) The sensitivity of individual organs to tumor induction in the fetus and in the child is not necessarily greater than in the adult and in some organs it seems as it benign tumors characteristically follow irradiation in childhood, whereas malignant tumors follow irradiation in adult life. The bone marrow of the child and of the fetus appear to be no more sensitive to the induction of myeloid leukemia than the bone marrow of the adult. (7) A formal scheme was presented by which dose limits for individual parts of the body might be determined. (8) The dose to the critical organ from any particular mode of radiation exposure does not define the overall risk, which will always be greater to the extent that other organs are irradiated. The concept of "critical" organs is administratively convenient, and in some circumstances logically justifiable, but it does not allow summation of risks according to the relative sensitivity of the irradiated tissue.

NCRP REPORTS 1966 TO 1971

Soon after the establishment of the Council under a Congressional Charter in 1964, the board of directors decided that the council would set up its own program for publication and distribution of reports and established a special office within the secretariat for this purpose. The main changes in the reports themselves were in the general format and the inclusion of a detailed index for each report.

The first report under the new procedure, No. 32, was on "Radiation Protection and Educational Institutions," developed under the Chairmanship of Dr. F. J. Shore and issued in July 1966.[90] It constituted the first report dealing with problems involved when radiation producing devices of our contemporary technology were used in the teaching of science in high school and undergraduate college levels. In particular, it sought to provide information on the hazards involved in the use of radiation producing equipment or radioactive materials in science demonstrations and experiments, and on the means of protection available to offset these hazards.

The report resulted from the original concern of the American Association of Physics Teachers (AAPT), which pointed out the potential hazards that exist in the indiscriminate use of radiation sources in education institutions and asked the Council to study the problem.

In addition to the support of the AAPT, the Public Health Service also assisted in the effort and during initial stages of the study made a determination of the scope and magnitude of the problem involved in current practices in educational institutions. To do this, the NCRP sent out a questionnaire circular to a number of schools and then the Public Health Service conducted a field survey in four of the schools which had responded meaningfully. Measurements were made and practices were evaluated. The survey developed significant information that the respondents to the NCRP questionnaire had not reported nor realized.

The report itself included a brief discussion on biological considerations, including somatic and genetic effects and the concept of maximum permissible dose, with the discussion directed to readers who were not likely to be too familiar with the radiation field. This was followed by a discussion of the typical radiation sources used in schools and the problems attendant on their use. It also gave some tabular information for radioactivity constants for different radionuclides and typical radiation levels around given kinds of equipment. Another chapter gave the basic principles of radiation protection drawn from other handbooks but presented in a relatively simplified form. A fifth chapter dealt with the elements of radiation protection programs that might be adapted to application in teaching institutions.

The most important new element introduced by this report were four new recommendations on maximum permissible doses for use in educational institutions, as follows:

1. Persons in the general population at any age — Such individuals should not receive an exposure exceeding 0.5 rem per year in addition to natural background and medical exposure. (This limit applies to those persons who were not occupationally exposed.) If a teacher or student of age 18 or greater is subjected routinely to work involving radiation as discussed in NCRP Report No. 17 (later superseded by Report No. 39)[98] then he is occupationally exposed and the exposure limits defined in those reports apply). Medical and dental exposures are to be made at the discretion of the doctor or dentist and are not to be considered in the evaluation of the individual's whole body exposure.

2. Persons under 18 years of age - These young people shall not be occupationally exposed to radiation (they should not be employed or trained in an x-ray department, radioisotope laboratory or industrial radiation facility).

3. Students under 18 years of age exposed during educational activities - Such individuals should not receive whole body exposure exceeding 0.1 rem per year *due to their education activity*. To provide an additional factor of safety, it is recommended that each experiment be so planned that no individual receives more than 0.01 rem while carrying it out.

**4. Students over 18 years of age exposed during educational activities fall into Category 1 above.

Report No. 33 on "Medical X-ray and Gamma Ray Protection for Energies up to 10 MeV (equipment design and use)," was prepared under the Chairmanship of Dr. R. O. Gorson and was issued in February 1968.[91] This was the first of two reports that were to be regarded as superseding the earlier Report No. 24 on the Use of Gamma Ray Sources, and Report No. 26 on the Use of X-ray Sources. Each of these reports had dealt with equipment design and use as well as structural shielding. However, it was decided that a more useful approach would result from the consideration of design and operational problems in one report, and structural shielding in another. Report No. 33 was concerned primarily with the design and operational aspects of medical x-ray equipment and gamma beam equipment.

It was emphasized in this report that its intention was to serve as a guide to good practice in medical radiation protection. While it provided basic standards for use in the preparation of regulatory protection codes, it was not specifically written for adoption as legal regulation. The report contained a number of recommendations concerning the design and performance characteristics of medical radiation-producing equipment and the manner in which it is used. The recommendations vary in importance and in applicability; some were particularly important for large busy installations but not for installations with very little workload; others would apply to all equipment of a given kind, whereas still others need not apply to equipment designed prior to the publication of the report. The Council expressed the belief that the risks involved in the judicious use of older equipment, failing to meet the revised standards of the new report, were not necessarily so great as to justify the condemnation of thousands of otherwise satisfactory units.

The report gave increased emphasis to problems associated with unnecessary patient exposure such as had been included in the various NCRP recommendations since 1931[55] and noted that there was still a substantial amount of loose practice in this regard. The PHS survey had shown that the genetically significant dose resulting from medical procedures in the United States was of the order of 55 millirems.[41] Analysis of these results also indicated that substantial reductions in this amount should be possible if greater attention were paid to a few simple precautions which had been frequently pointed out in earlier NCRP reports as well as reports of other groups.

The precautions were summarized in a section on the general guidelines in the clinical use of radiation and were then enlarged in the detailed discussions throughout the report. The most important recommendation was that the useful beam should be limited to the smallest area practicable and should be consistent with the objectives of the radiological examination or treatment.

The voltage, filtration, and source-skin distance (SSD) employed in medical radiological examinations should be as great as is practical and should be consistent with the diagnostic objectives of the study.

Protection of the embryo or fetus during radiological examination or treatment of women known to be pregnant should be given special consideration. Ideally, abdominal radiological examinations of a woman of childbearing age should be performed during the first few days following the onset of menses to minimize the possibility of radiation of an embryo. In practice, medical needs should be the primary factor in deciding the timing of the examination.

Suitable protective devices to shield the gonads of patients who are potentially procreative should be used when the examination or method of treatment may include the gonads in the useful beam, unless such devices interfere with the conditions or objectives of the examination or treatment.

Fluoroscopy should not be used as a substitute for radiography but should be reserved for the study of dynamics or spatial relationships or for guidance in spot-film recording of critical details.

X-ray film intensifying screens and other image recording devices should be as sensitive as is consistent with the requirements of the examination.

Film processing materials and techniques should be those recommended by the x-ray film manufacturer or those otherwise tested to insure maximum information content of the developed x-ray film and, where practical, quality control methods should be employed to insure optimum results.

Except for the increased emphasis on exposure of women and shielding of the gonads, all of these points had been previously recommended for many years.

The remainder of the report was divided into chapters on X-ray Equipment, Gamma-Beam Therapy Equipment, Therapy Equipment Calibration Guide, Radiation Protection Surveys, and Working Conditions.

An appendix gave an example of the kind of emergency procedures that should be established for situations where there may be a failure in the gamma-beam control mechanisms. This was presented as sample rather than as a specific recommendation.

The second report of this pair, entitled "Medical X-ray and Gamma Ray Protection for Energies up to 10MeV (Structural Shielding Design and Evaluation)" (Report No. 34) was prepared under the Chairmanship of C. B. Braestrup and was issued in March 1970.[93] As evident from the title, and as explained above, this report was devoted exclusively to shielding design information and the necessary data for architectural and engineering use. While it provided basic standards in this area which might be used in the preparation of regulatory protection codes, it was not specifically written for adoption as legal regulation.

In past reports by the NCRP, problems concerning protection in the dental use of x-rays was embodied in the same report as medical x-rays. However, it was decided by the Council to formulate one report dealing primarily with the dental problem, embodying both the design and operational aspects of dental x-ray equipment and matters relating to structural shielding design. It was estimated that there might be some hundred thousand dental units in the country and that in general the literature sources for the dentist and the radiologist did not adequately coincide.

To meet this special need a report on "Dental X-ray Protection" (Report No. 35) was prepared under the Chairmanship of Dr. R. J. Nelsen and issued in March 1970.[94]

The sections on Equipment, Shielding Design, Protection Surveys, and Working Conditions pretty much follow the earlier reports and general recommendations; however, the problem of operating procedures for the dentist, as compared to the physician, may be quite different. Special emphasis was laid on reduction of unnecessary exposure of the patient.

Reference was made to the use of film holders locked to the x-ray tube in such a manner that positive alignment was always assured. After considerable examination of this problem it was decided that there could not be a blanket recommendation requiring such devices but the report indicated that there were such situations where they could be useful if evaluated for the particular circumstances.

A second area in which there had been considerable public agitation had to do with gonad shielding; again the report indicated that unless direct exposure of the gonads to the useful beam might occur, the use of a lead rubber apron was not indicated. Surveys had shown that in proper dental radiography, gonadal exposure of the patient is due almost always to the scattered radiation alone, and that in comparison with other normal radiation exposures its practical significance makes the further reduction relatively unimportant.[41] However, when unconventional projections are used, and the gonads may be in the useful beam, gonadal shielding was indicated as mandatory.

As in earlier reports, emphasis was again put on the need for careful beam alignment, proper film selection, and proper processing of exposed films.

Another of the new series of reports, "Radiation Protection in Veterinary Medicine," (Report No. 36), was prepared under the Chairmanship of Dr. B. F. Trum and issued in August 1970.[95] Here, again, much of the basic information is the same or similar to that in the medical protection field. However, there are certain distinctly different problems in dealing with animals as compared to dealing with people. For example, there does not need to be the same great concern about hereditary effects or the sharp limitation of "patient" exposure. Also, there are distinctly different equipment requirements and procedures in handling the equipment.

While it is recommended that no individual shall be used routinely to support or hold animals or film during exposures, there are sometimes conditions such that this would be the only possible way of achieving the desired result. Accordingly, it was specified that if an animal patient had to be held or positioned manually, the individual doing this should wear protective gloves and protective aprons. Caution was also issued against the use of pregnant or potentially pregnant women and individuals under 18 to support or hold animals during such exposure.

Except for such examples as noted above, the report follows fairly closely the established practice for dealing with protection in either the diagnosis or treatment of patients. However, so as to make the recommendations more readily accessible to the veterinary specialist, the material was assembled and fitted to suit his needs.

A new report on "Precautions in the Management of Patients Who Have Received Therapeutic Amounts of Radionuclides" (Report No. 37) was prepared under the Chairmanship of Dr. Edith H. Quimby and issued in October 1970.[96] This was the third of a series of reports dealing with patients containing radionuclides. However, the earlier ones were devoted more specifically to the handling of cadavers, whereas the new one put new emphasis on the problems centering about living patients.

The report included updated information and recommendations with regard to preparation for burial or cremation, with or without autopsy, and for the handling of accident or injury during either surgery or autopsy. New sections were introduced on Hospital Routine and Nursing Care for Radioactive Patients, providing information on exposure rates at different distances from the patients for the more commonly used radionuclides. In this, the special section dealt with treatment and with encapsulated sources whether premanent or removable, but which are mechanically inserted, and also treatment solutions, collodial suspensions, or microspheres. Recommendations were given for the protection of other patients and visitors providing some guidance as to the distance which visitors should remain from the patient together with length of visit, etc.

A new section dealt with the Release from Hospitals of Patients Containing Radioactive Material, and for this purpose some new recommendations were given for the first time. Recognizing that there may be some relatively rare and unusual situations where it would be necessary or highly desirable to send a patient home, in spite of his carrying a burden that could result in a dose to other persons in excess of 0.5 rem, it was recommended that such cases be permitted, *as exceptions*, provided in general that (1) no person under the age of 45 years shall be permitted to receive a dose in excess of 0.5 rem in a year, (2) no person over the age of 45 years shall be permitted to receive a dose in excess of 5 rems in a year, (3) the circumstances leading to the decision to make an exception, the evaluation of the exposure conditions, and the means of controlling individual exposure shall be documented and, (4) local health authorities shall be notified of the action. The four recommendations above were subsequently included as a part of the regular basic radiation protection criteria.

Detailed information and recommendations were given relative to the discharge of patients from the hospital, the return of the patient to work, and actions to be taken in the case of the death of a radioactive patient at home.

New recommendations were presented covering emergency surgery or death of the radioactive patient, the handling of contaminated clothing or instruments, and cremation.

Several typical radioactivity tags, labels for the patient's chart, his room, etc. were given.

REPORT ON NEUTRON PROTECTION

A new report entitled "Protection Against Neutron Radiation" (Report No. 38) was developed under the Chairmanship of Dr. H. H. Rossi and issued in January 1971.[97] This was an updating and extension of the 1957 report on the same subject.[75] In the years following the issuance of the earlier report, a number of developments had taken place making it desirable to issue a new one. Most of the changes had been prompted by a new formulation of applicable quantities and units carried out by the ICRU and changes in the maximum permissible dose equivalents recommended by the NCRP and the ICRP.

After a section on the General Principles of Neutron Protection, the basic elements of neutron protection were developed. This included classification of the neutrons and primary modes of interaction, radiation quantity, radiation quality, and the interactions between neutrons and tissues. A new table of mean quality factors (QF) was given for a wide range of neutron energies and flux densities.

A section was included on the hazard of exposure to combined x- and gamma rays arising in the operation of neutron sources.

A chapter on Radiation Protection in the Installation and Operation of Neutron Sources included recommendations for various types of neutron sources, shields, signs and barriers, and procedures in case of significant overexposure.

A section gave a set of rules pertaining to maximum permissible dose equivalent. These rules were based on some of the concepts originally developed in Report No. 17,[71] and the modifications of 1956.[77] For example, they included the limitations based on a dose in any 13 consecutive weeks, a concept which has since been removed from the NCRP recommendation. Similarly, the new levels for the forearms, feet, and ankles should be considered as obsolete. Although these older concepts were included in the report, and some of the tabular information relates to them, the material is still completely usable in application to the new levels that were put forth in Report No. 39 particularly for occupational exposure.

The report included nearly 100 pages of appendices including tables and curves on depth dose, reactions employed in neutron production, neutron-capture gamma rays, shielding data, and accident dosimetry. Most of these represent the latest and most up-to-date technical information and a decided technical improvement over the earlier report.

NCRP REVIEW OF BASIC RADIATION PROTECTION CRITERIA (1959-1971)

The last comprehensive report on Basic Radiation Protection Standards and Criteria was completed by the NCRP in 1949 and published in 1954.[71] In 1957 the maximum permissible dose (MPD) was reduced to 5 rem per year,[74] and in 1959 the work by the Ad Hoc Committee on Somatic Dose for the Population was released.[79] Beginning at that time, the original committee 1 on Permissible Dose from External Sources of Ionizing Radiation was regrouped under the Chairmanship of Dr. S. Cantril who, unfortunately, died a few years later, at which time H. M. Parker was made the new chairman. During the earlier period the Committee had attempted primarily to modify the original Report No. 17 by essentially inserting the new numbers resulting from the 1957 and 1959 modifications. After substantial effort had been devoted to this, it became evident that there had been new developments followed by changes in other concepts, and it was decided to start a completely new presentation.

Throughout the 1960's the Committee, at times assisted by the Executive Committee or the Board of Directors, carried out a continuing review of the literature and new developments. One of the most important of these was the genetics work of Russell, first reported in 1958 and then successively strengthened by his own laboratory and others, as already noted above. Also during this period, fruitful results began to develop from the studies of the Atomic Bomb Casualty Commission, specifically establishing a significant relationship between the dose to survivors of the Atomic Bomb in Japan and the incidence of leukemia, thyroid cancer, cataracts, etc. During this same period, accented by the joint studies of the ICRP and the ICRU, increased attention was directed to some of the problems of radiation exposure during medical procedures.[106,110] Also during this period, studies of the effects of radiation on the fetus began to be reported by Court-Brown and Doll and Alice Stewart in England, as well as other studies abroad, and the studies by MacMahon and others in this country. While the full interpretation of these data is still incomplete, there seemed to be sufficient evidence for added caution in fetal exposure. At the same time, through the continued evaluation, it became increasingly evident that the general philosophy and reasoning developed in the earlier report would remain essentially unchanged except for updating with such new information as referred to above. Even taking the new information into account, the changes suggested were primarily in detail and certainly not in overall basic concept.

The new report entitled "Basic Radiation Protection Criteria" was completed in April 1969,[98] but because of continued review by all of the members of the Council, and because of some proposals being made by non-members of the Council, its issuance was delayed until January 1971. Throughout the period since the first report was issued, there was increasing concern on the part of the Council about the growing complexity in the numerical expression of the standards that seemed to imply a better basic knowledge about the subject than actually existed. As a consequence, a special endeavor was made to somewhat simplify the overall numerical structure and at the same time to make small adjustments in some of the secondary standards wherever current technology or need indicated. Considerable use was made of the extensive discussions and data sources that had been developed by the ICRP, and ICRU, UNSCEAR, the National Academy of Sciences, the British Medical Research Council, the Federal Radiation Council, and other such groups. A brief summary of the principal recommendations in the 1971 report follows:

The annual *prospective* maximum permissible dose equivalent to the critical organs for radiation workers will remain at 5 rems per year. It was noted that while this has been a permissible amount for some 13 or 14 years, only a very small percentage of radiation workers ever receive more than about 1 rem per year.

Retrospectively exposures as high as 10 or even 15 rems per year would not be considered to be of medical significance, but might well have implications as to protection practices within the particular installation. Allowance of any such levels of exposure would depend upon the accumulation rate and be subject to the 5 (N-18) restriction.

Exposure of the hands would be limited to 75 rem per year at a rate no greater than 25 rems per quarter. It was noted that the original intention of the Committee was to set this level substantially lower. It had been introduced originally because it was felt to meet the needs of medical practice, but the Committee was surprised to find at this time that reducing it any further would introduce substantial difficulties in industry and in the meantime there was no strong evidence that the older, higher levels had caused difficulty.

Exposure of the forearms is no longer considered in the limitation for the hands. For the forearms a level of 30 rems per year at a rate not greater than 10 rems per quarter has been recommended.

All tissues and organs, other than those defined as critical, (blood forming organs, gonads, and eyes) have been set at 15 rems per year. The dose resulting from a specified body burden of radium has been dropped as a base line for internal emitters, but remains a useful measurement. Here it will also be noted, especially in view of some of the recent observations on the Marshallese, that there seemed to be no reason or need for keeping the permissible dose to the thyroid at double that set for other organs.

Dose limits for the population remain unchanged at 0.5 rem per year for the individual, with 0.1 rem of this permitted for students. The currently accepted value of 0.17 rem per year for an average population exposure was retained but it might be noted here that this number had not been stated unequivocally in the past by the NCRP.

Accidental and emergency exposure limits were left the same as originally set in 1949 and the report carefully avoided any formulation of procedures to be followed in the event of occasional overruns. It was felt that regulation on such matters has no biomedical significance and could even handicap the worker or the organization, or both, due to lack of flexibility. Experience has proven that such situations are rare and actions regarding future work of a person so overexposed will be based on a case-by-case medical judgment.

The report repeatedly emphasized that where some changes were made in permissible dose levels or dose limits, these were primarily for the purpose of developing a less complex numerical system of standards. There was, in fact, very little, if any, evidence indicating that the earlier standards would lead to unacceptable exposures.

Some general guidance, but definitely not permissible levels, have been outlined for severe emergencies, possibly involving the saving of lives. Upper limits were suggested at 100 rems to the critical organs (blood forming, gonads, eyes) with the possibility of 200 rems additional to the hands and forearms. To the extent possible, persons older than 45 should be used in any such planned operation.

For less urgent emergencies, levels of 25 and 100 rems, respectively, were suggested.

In recognition of the problem of patients containing radioactivity and remaining at home, dose limits for the family have been set at the individual population level of 0.5 rem per year for persons under age 45 and 5 rems per year for individuals over age 45. (See discussion in Report No. 37 above.)[9 6]

The report recognized, especially in the development of dose limits such as just indicated, that some of these recommendations may present management difficulties. But it is precisely for that reason that it was felt that some guidance would be useful because of the experience in the United States that when such guidance was unavailable, regulatory restrictions tended to become unduly severe.

The Scientific Committee also directed considerable attention to the possible interrelationship of necessary diagnostic or therapeutic exposures to the occupational exposure of radiation workers. It could find no basis, under most circumstances, for applying any radiation work restrictions on individuals because of ordinary diagnostic x-ray exposures, or even of most therapeutic exposures. The increment of either, in comparison with the other, is unimportant. There appears to be no reason why a person who has been subjected to a medical irradiation should be removed from normal radiation work, or why a radiation worker should be restricted as to the number of proper medical x-ray exposures that he might be required to receive. The principal exception to this would be cases of where one person had medical doses running into the 10's or 100's of rads and where, in his radiation work he might be subjected to large accidental overexposure.

The report was divided into eight chapters. Chapter I, a discussion of radiation sources, contained a general discussion about natural radiation and the normal uses of radiation as an important resource for man. Chapter II dealt with radiation exposure conditions that might require consideration, such as natural radiation, radiation from medical procedures, occupational irradiation, and man-made environmental radiation. Conditions of radiation control were also covered to include such things as controlled areas and the control of individuals entering areas where radiation exposure might occur.

Chapter III dealt with a broad discussion of the basic biological factors; Chapter IV with radiation effects, including genetic effects, neoplastic diseases, cataracts, growth and development, and life span. Chapter V dealt with the manifestations of overexposure in adults. This was primarily to give some general idea of the effects of relatively large exposures such as might be encountered in major nuclear accidents, nuclear warfare, etc.

The second part of the report dealt with Radiation Protection Standards, starting in Chapter VI with the general bases used for radiation standards. This included discussions of risk, terminology, the legitimacy of varying standards for varying situations, dose allocations for the population, and the significance of standards as administrative guides or controls for occupational exposure. Chapter VII dealt with specific Radiation Protection Concepts or Standards dealing with critical organs and tissues and specification of the dose received. The final Chapter VIII gave the Dose Limiting Recommendations, derived from the preceding discussions, and provided some guidance for special cases such as already discussed above.

As a part of the discussion on the basic problem of radiation effects and risk concept, there was an explanation of some of the concepts upon which subjective judgment had to be made. These included the problem of extrapolation from high level effects, which could be observed quantitatively in relation to dose, to effects that might be postulated in the low dose region for which no effects have ever been observed. There was also a detailed discussion of the concept of threshold versus non-threshold effects of radiation and the influence of the assumption of the non-threshold concept upon the development of standards.

Table 1 gives a summary of the 1971 recommendations of the NCRP, but the reader should be cautioned that the application of dose limits is substantially conditioned by the qualifications and comments provided in the original report.[98]

POLITICAL ACTION ACTIVITIES

Another dimension has been added to the problem of the development and acceptance of radiation protection standards, especially in, but not limited to, the United States. As already noted in several places above, all responsible bodies devoted to the development of radiation protection standards have for many years recognized the

TABLE 1

1971 Dose-Limiting Recommendations (NCRP)

Maximum Permissible Dose Equivalent for	
Occupational Exposure	
combined whole body occupational exposure	
Prospective annual limit	5 rems in any one year
Retrospective annual limit	10 - 15 rems in any one year
Long-term accumulation to age N years	(n - 18) X 5 rems
Skin	15 rems in any one year
Hands	75 rems in any one year (25/qtr)
Forearms	30 rems in any one year (10/qtr)
Other organs, tissues, and organ systems	15 rems in any one year (5 qtr)
Fertile women (with respect to fetus)	0.5 rem in gestation period
Dose Limits for the Public, or Occasionally	
Exposur Exposed Individuals	0.5 rem in any one year
Individual or Occasional Students	0.1 rem in any one year
Population Dose Limits	
Genetic	0.17 rem average per year
Somatic	0.17 rem average per year
Emergency Dose Limits – Life Saving	
Individual (older than 45 years if possible)	100 rems
Hands and Forearms	200 rems, additional (300 rems, total)
Emergency Dose Limits – Less Urgent	
Individual	25 rems
Hands and Forearms	100 rems, total
Family of Radioactive Patients	
Individual (under age 45)	0.5 rem in any one year
Individual (over age 45)	5 rems in any one year

difficulty of setting standards in the very low dose region on a sound scientific and technical basis. As an assistance toward this end, theories and postulates have been adopted which, if interpreted literally, would lead to the conclusion that any amount of radiation, however, small, may lead to some possibility of fatal effects. At the same time, it has been generally recognized by radiation protection bodies that while these theories and assumptions form a reasonable baseline for evaluation of radiation risks, there are known deviations from them in the region of low doses and doses delivered over long periods of time.

In spite of this, a few people have treated the theories and postulates as facts and have gone to the public and to the Congressional Committees, Reactor Hearings, and the Courts in opposition to nearly any activity which may develop radiation. This has resulted in a great deal of confusion and apprehension in the public and has been especially misleading to the legislatures, hearing boards, and agencies charged with promoting radiation health safety.

Since, in general, the representations by these individuals are not regarded as scientifically acceptable they will not be discussed or refuted here. Nevertheless, they must be recognized as a problem which must be confronted in the development of radiation practices, probably for some few years to come; because of the possibility of their importantly influencing the specification of radiation standards, they will remain a problem to be reckoned with.

UNITED NATIONS SCIENTIFIC COMMITTEE ON THE EFFECTS OF ATOMIC RADIATION (UNSCEAR)

As already noted, UNSCEAR was organized in October 1956, and has issued four major reports. In addition to the arrangement of the material and the observations and conclusions which they draw, these reports are veritable gold mines of information and bibliographic referencing. A few of the highlights from each of the reports will be outlined below. While many of the points and issues which they discuss have already been brought out to some degree, either by the NCRP or the ICRP, there is added significance to many of them because they have been equally recognized and developed by a scientific committee operating at the international political level. It can almost be said that when data are analyzed, conflicting opinions resolved, and conclusions drawn by the UNSCEAR, the procedure can be said to have come very close to the adversary method of resolving points of issue. When agreement is reached between nations having widely different scientific and political approaches to problems, the final conclusions may be somewhat thinned or watered down, but any points of agreement are likely to have greater strength simply by reason of the way at which they were arrived.

The first report (Supplement No. 17, A/3838) was issued in 1958.[1,2,3] It should be borne in mind that UNSCEAR came into being largely because of the international problem of worldwide fallout from weapons testing by various countries and undoubtedly their original intention was to evaluate this in terms of both health and political impact. As a consequence, a great deal of material on all kinds of radiation exposures was collected, evaluated, and tabulated as supplements to their main discussion, summary, and conclusions. The final conclusions represented 62 items of which 53 were regarded as special, and the remainder general, in conclusion.

Of the major sources of radiation they noted that radiation from natural sources varied widely according to geographical location and that occasional populated areas exceeded the average by a factor of 10. Exposure due to medical procedures was greater than through any other class of source. Occupational exposure constituted only a small fraction of the total radiation of the population, amounting to about 2% of that from natural sources in countries in which the occupational exposure was probably the largest.

It appeared that the discharge of radioactive wastes had not led to appreciable radiation exposure of populations but because of the possible greater use of reactors, they felt that this kind of source should be watched closely. Radiation from fallout resulting from weapons tests was very small but might persist for long periods of time.

Recognizing the complexity of the evaluation of the biological effects of radiation, cautious statements were made on the effects of low level exposures over long periods of time. They recognized the possibility that embryonic cells might be especially sensitive to radiation and noted that some evidence suggested that exposure of the fetus to small doses of radiation could result in leukemia during childhood.

Children were regarded as being more sensitive to radiation than adults, although there was little direct evidence on the subject except for an indication that cancer of the thyroid might result from doses of a few hundred rads which do not induce this change in adults.

In human adults they found it was difficult to detect the effect of a single exposure of less than 25 to 50 rem, or of continuing exposures to levels below 100 times the natural levels. Transient disturbances were reported to have been observed in mammals after exposure to a single dose of 25 to 200 millirem.

The processes of repair were recognized as probably playing an important role in the final outcome of radiation damage. They noted that some thresholds were found for somatic effects such as erythema of the skin but other forms of radiation damage to cells, tissues, or organisms, appeared to be cumulative; for instance, mutational damage, once established, is not repaired. (See their later statement on this subject.)

They noted that damaged cells or tissues may be eliminated and replaced by regenerated normal cells, this process being most active in embryos and young animals and in certain tissues of the adult. The power of repair differs considerably in different organisms any type of cells and varies to a high degree with physiological conditions.

They expressed the opinion that exposure of gonads to even the smallest doses of ionizing radiations can give rise to mutagens which accumulate, are transmissible to the progeny, and are in general, considered to be harmful to the human race. The present assumption of the strictly accumulative affect of radiation in inducing mutations in man was based upon theoretical considerations and a limited amount of experimental data obtained by the exposure of experimental organisms to relatively high dose levels.

With emphasis, they stated, "Any present attempt to evaluate the effects of sources of radiation to which the world population is exposed, can produce only tentative estimates with wide margins of uncertainty."

The general summary concluded with a number of items giving positive indications for new research.

The second report (Supplement No. 16 A/5216) was issued in 1962.[124] and dealt in considerably greater depth with the physical and biological aspects of the interaction of ionizing radiation with matter, somatic effects, hereditary effects, sources of irradiation, and comparison of doses and estimates of risks.

The general evaluations and conclusions followed somewhat the pattern developed earlier but they were presented in a more quantitative form; for example, the average radiation dose from natural sources was estimated for various tissues as about 125 mrem to the gonads, 120 mrem to the blood forming cells, and 130 mrem to the cells lining the bone surfaces.

Estimates of the annual genetically significant dose received from medical procedures ranged from 6 to 60 mrem in those countries in which surveys had been carried out. Tentative estimates for bone marrow doses ranged from 50 to 100 mrem for the yearly contribution. Average occupational exposures in four countries led to estimates of less than 0.5 mrem per year.

In discussing the somatic effects, it was noted that during the interval since their last report, the knowledge of the late effects of radiation on man had increased substantially with the demonstration of the induction of certain transient somatic effects by low doses of a few rads of radiation and with the confirmation that embryonic tissues are more sensitive to injury by radiation than many adult ones.

They noted, as they had earlier, that radiation exposure of animals continued for long or short periods, causes a shortening of the life span by an amount depending upon the dose received and the dose rate. They noted the probability that a similar life shortening occurs in man but that the evidence on this point is inconclusive and no estimate could be given of the amount of any such effect.

In dealing with the question of hereditary effects, they noted that in human genetics there had been significant progress in the research since 1958 and that an entirely new field of study had been opened up owing to recent cytogenetic findings in man. The concept of mutation induction as an instantaneous process was revised and evidence has accumulated showing that for some mutations a finite period of time elapses between the absorption of radiant energy and the completion of the mutation process during which, depending on the physiological state of the cell, at least partial repair of the damage may be possible.

The frequency of gene mutations produced by a radiation was noted to have been proportional to the total dose received by the germ cells. However, for mice, fruit flies, and silkworms, the proportionality has been shown to vary with certain factors including dose rate.[47] The dose required to induce as many mutations as occur naturally, the so-called "doubling dose," therefore also changes with the dose rate. (These later findings were noted as being a profound change from the theories that had been held before relative to the genetic effects of radiation. Applied to the development of radiation protection standards they clearly imply a relaxation rather than a tightening.)

In their general conclusions, while still holding to the concept that exposure to radiation, even in doses substantially lower than those producing accute effects, may occasionally give rise to a wide variety of harmful effects including cancer, leukemia, and inherited abnormalities, these might not be readily distinguishable from naturally occurring conditions or identifiable as due to radiation. This led to their main conclusion that the need for all forms of unnecessary radiation exposure should be minimized or avoided entirely, particularly when exposure of large populations is entailed; and that every procedure involving the peaceful uses of ionizing radiation should be subject to appropriate immediate and continuing scrutiny to ensure that the resulting exposure is kept to the minimum practicable level and that this level is consistent with the necessity or the value of the procedure.

The third UNSCEAR Report (Supplement No. 14, A/5814) was issued in 1964.[125] In this it was noted that the report made no attempt to cover the whole field of radiation effects as had the 1958 and 1962 reports of the Committee. Instead it confined itself to two principal subjects: the contamination of the environment by nuclear explosions and the possibility of quantitatively assessing the risk of induction of malignancies by radiation in man.

As a consequence of these objectives, the report consisted in the main of the compilation and evaluation of much of the technical material that was considered. It carried a bibliography of 138 items in addition to some 200 documents submitted by various participating countries.

The fourth Report (Supplement No. 14, A/6314) was issued in 1966 and dealt with two principal items:[126] environmental radiation and the genetic risks of ionizing radiation. Similar to the previous report, the bulk of this one was devoted to data review and evaluation. Their conclusions regarding the genetic effects of radiation followed along the pattern that had begun to develop during the preceding five years, but in this report they were stated with much greater confidence. They noted that a new estimate has been obtained for the spontaneous frequency of gene mutations over the whole of the hereditary material of man. An estimate was made of the rate of induction of gene mutations per unit of radiation dose. From these it would appear that a dose of 1 rad per generation would add something like 1/70th to the total number of mutations arising spontaneously in a generation.

The proportion of 1/70th might also apply to hereditary diseases of man which are known to be important and which can be transmitted directly from parent to offspring, but it was emphasized once more that these diseases contribute only a small proportion to the damage from gene mutations. They noted some evidence that complex inherited characteristics, such as stature and intelligence, might be affected by induced gene mutations and that the effects would probably be adverse.

The opinion was expressed that one fourth of all abortions are caused by, and 1% of all live-born infants suffer from, severe effects of chromosomal abnormalities which arise spontaneously. It was also indicated that in our present state of knowledge it is only possible to give estimates of rates of induction by high doses of radiation of chromosomal damage of types which include not more than a small proportion of abnormalities that occur naturally. The number of these that would arise after exposure to high doses can be estimated, but it is not known how many would occur following low doses, although the yield per unit dose would be much less than expected if the yield were directly proportional to the dose; a large part of this type of genetic damage would not be expected to persist in a population for more than one generation. In this connection evidence was presented that from experience using high doses, it is known that malformations of the skeleton do occur fairly frequently in these offspring, but that whether proportional numbers of such defects would result from low doses to parents is not known.

The report emphasized that the estimates arrived at relate to the genetic effects of acute exposures at high doses to male reproductive cells in the stage (spermatogonia) that is most important in human hazards. Lower numbers of mutations per unit doze will occur where the radiation dose is low or spread out over a long time. It was also known that the reproductive cells of the two sexes differ in sensitivity; fewer mutations, on the average, will occur when the reproductive cells of the females (oöcytes) are exposed to radiation. The Committee concluded that because all of the estimates that they had noted were subject to so many uncertainties, they should not be applied in a single simple and direct fashion to radiation protection.

The review above had had to be somewhat channeled and this has been done with accent on radiation protection developments in the United States, and those elements that have contributed most directly to our current status. Many areas have been passed by, to be treated later. These will include such items as: National radiation control experience in other countries such as United Kingdom and Sweden; International Atomic Energy Agency; United States of America Standards Institute; International Standards Organization; more on the National Council on Radiation Protection and the International Commission on Radiological Protection; more on the Federal Radiation Council, Public Health Service/Bureau of Radiological Health; Radiation Control for Health and Safety Act and related actions, etc. The writer hopes to add these elements as time permits.

TABLE 2

Some Important Dates in the Development of Radiation Protection Standards

Year	Event
1915	British Roentgen Society proposals for radiation protection.
1921	British adopt radiation protection recommendations.
1922	American Roentgen Ray Society adopts radiation protection rules.
1928	Unit of x-ray intensity proposed by Second International Congress of Radiology.
1928	International Committee on X-ray and Radium Protection established.
1928	First international recommendations on radiation protection adopted by Second International Congress of Radiology.
1929	Advisory Committee on X-ray and Radium Protection established (U.S.).
1931	The Roentgen adopted as a unit of x radiation.
1934	Tolerance dose of 0.1 R/day recommended by Advisory Committee on X-ray and Radium Protection (March).
1934	Tolerance dose on 0.2 R/day recommended by International Committee on X-ray and Radium Protection (July).
1941	Advisory Committee on X-ray and Radium Protection recommends 0.1 μ Ci permissible body burden for radium.
1946	Advisory Committee on X-ray and Radium Protection reorganized to the National Committee on Radiation Protection.
1949	National Committee on Radiation Protection lowers basic MPD for radiation workers to 0.3 rem/week. Risk-benefit philosophy introduced.
1950	International Commission on Radiological Protection and International Commission on Radiological Units reorganized from pre-war dommittees.
1950	International Commission on Radiological Protection adopts basic MPD of 0.3 rem/week for radiation workers.
1953	International Commission on Radiological Units introduces concept of absorbed dose.
1956	National Academy of Sciences and International Commission on Radiological Protection recommends lower basic permissible dose for radiation workers — 5 4rem/yer.
1957	National Committee on Radiation Protection and Measurements introduces age proration concept for occupational exposure and 0.5 rem/year for individuals in population.

Table 2 (continued)

1959	International Commission on Radiological Protection recommends limitation of genetically significant dose to population of 5 rems in 30 years.
1964	Federal Radiation Council introduces concept of protective action guides.
1971	National Council on Radiation Protection and Measurements recommends same value of 15 rem/year for all non-critical organs.

REFERENCES

1. Kassabian, M. K., *Electrotherapeutics and Roentgen-Rays*, J. B. Lippincott Co., Philadelphia, 1907.

2. Grubbé, E. H., *X-ray Treatment, Its Origin, Birth and Early History*, Bruce Pub. Co., St. Paul, Minn., 1949.

3. Edison, T. H. et al., Effect of x-rays upon the eye, *Nature*, 53, 421, 1896.

4. Tesla, N., X-rays, *Electrical Rev.*, December 2, 1896.

5. Glasser, Otto, First observations on the physiological effects of roentgen rays in the human skin, *Amer. Roentgen.*, 28, 75, 1932

6. Rollins, W., The control guinea pigs, *Boston Med. Surg. J.* February 21, 28, and March 28, 1901.

7. Rollins, W., Vacuum Tube Burns, *Boston Med. Surg. J.* January 9, 1902

8. Roentgen, W. C., On a new kind of rays, second communication, *Sitz. Phys. Med. Geo.*, Würzburg, March 9, 1896. (*Ann. Phys.*, 64, 1898).

9. Rollins W., *Notes on X-Light*, Boston, 1903, 290.

10. International X-ray Unit Committee, International x-ray unit of intensity, *Brit. J. Radiol.* (N.S.), 1, 363, 1928.

11. Rollins, W., Non-radiable cases for x-light tubes, *Electrical Rev.*, June 14, 1902.

12. Russ, S., Hard or soft x-rays, *Brit. J. Radiol.*, 11, 110, 1915.

13. Kaye, G.W.C., *Xrays,* 2nd ed., Longmans, Green, London, 1915, 264.

14. Proceedings, Annual Meeting of American Roentgen Ray Society, September 1920 (L.S.T. Historical Files. See also **Protection in radiology, by Pfahler, G. E.,** *Amer. J. Roentgen.,* **9**, 803, 1922.)

15. **Recommendations of the British X-ray and Radium Protection Committee,** *Brit. J. Radiol.,* 17, 3, 50, 52, 53, 98, 100, 1921.

16. Kaye, G. W. C., *X-rays* 4th ed. 289, Longmans, Green, London, 1926.

17. Kaye, G. W. C., *Roentgenology,* 116, P. Hoeber, 1928.

18. Wintz, H. and Rump, W., Protective measures against dangers resulting from the use of radium, roentgen, and ultra violet rays, League of Nations, C. H. 1054, Geneva, 1931.

19. Mutscheller, A., Physical standards of protection against roentgen ray dangers, *Amer. J. Roentgen.,* 13, 65, 1925.

20. Kustner, N., *Strahlenther,* 26, 120, 1927.

21. Glocker, R. and Reuss, A., Strahlenschutzmessungen, *Fortschr. Geb. Roentgenstr.,* 40, 501, 1929.

22. Behnken, H., Dosimetrische Untersuchungen uber Roentgenstrahlenschutz und strahlenschutzrohren, *Fortschr. Geb. Roentgenstr.,* 41, 245, 1930.

23. Bouwers, A., and van der Tuuk, J. H., Strahlenschutz, *Fortschr. Geb. Roentgenstr.,* 41, 767, 1930.

24. Barclay A. E. and Cox, S., Radiation risks of the roentgenologist. *Amer. J. Roentgen.,* 19, 551, 1928.

25. Chantraine, H. and Profitlich, P., uber den schutz den Artzes und seiner gehilfen vor den Roentgenstrahlen, *Fortschr. Geb. Roentgenstr.,* 38, 121, 1928.

26. Solomon, I., Recherches sur la valeur des moyens de protection contre l'action a distance de rayons de roentgen, *J. Radiol.,* 8, 62, 1924.

27. Schechtmann, J., Die anwendung des aspirationsgerates nach Ebert zur prüfung des roentgenschutzes in den roentgenabteilungen, *Fortschr. Geb. Roentgenstr.,* 42, 645, 1930.

28. Fricke, H. and Beasley, I. E., Measurement of the stray radiation in roentgen-ray clinics, *Amer. J. Roentgen,* 18, 146, 1927.

29. Jacobson, L. E., Measurement of stray radiation in roentgen ray clinics of New York City, *Amer. J. Roentgen.,* 18, 149, 1927.

30. Sievert, R., The International Commission on radiological protection, International Associations, 9, 589, September 1957.

31. Taylor, L. S., History of the International Commission on radiological protection, *Health Phys.,* 1, 97, 1958.

32. International X-ray and Radium Protection Committee, X-ray and radium protection, *Brit. J. Radiol.,* (N.S.), 1, 358, 1928.

33. Stone, R.S., The concepts of maximum permissible exposure, *Radiology,* 58, 639, 1952.

34. Stone, R.S., Health protection activities of the plutonium project. *Proc. Amer. Phil. Soc.,* 90, 11, 1946.

35. Taylor, L.S., Brief history of National Committee on Radiation Protection, *Health Phys.,* 1, 3, 1958.

36. Fano, U. and Taylor, L.S., Dosage units for high energy radiation, *Radiology,* 55, 743, 1950.

37. Recommendations of the International Commission on Radiological Units, *Amer. J. Roentgen.,* 65, 99, 1957.

38. Conference on Radiobiology and Radiation Protection, Stockholm, September 15, 1952, *Acta. Radiol.,* 41, 1, 1954.

39. Sievert, R. M., Act concerning supervision of radiologic work (SFS No. 334/1941 and 267/1945), *Acta Radiol.,* 33, 206, 1950.

40. Medical Research Council, The hazards to man of nuclear and allied radiations, Cmd 9780, H. M. S. Office, 1956.

41. U.S. Public Health Service, Population dose from x-rays, PHS Pub., 2001, 1964.

42. Taylor, L. S., The philosophy underlying radiation protection, *Amer. J. Roentgen.,* 77, 914, 1957.

43. Taylor, L. S., Radiation exposure and the use of isotopes, *Impact*, 7, 209, 1957.

44. Bond, V. P. et al., Dose-effect modifying factors in radiation protection (Report of NCRP Com. M-4) BNL 50073 (T-471), Brookhaven Nat. Lab., 1967.

45. Braestrup, C. B. and Mooney, R. T., X-ray emission from television sets, *Science*, 130, 1071, 1959.

46. Executive Order 10831, August 14, 1959 and P. L. 86-373.

47. Russell, W. L., Russell, L. B., and Kelly, E. M., Radiation dose rate and mutation frequency, *Science*, 128, 1546, 1958.

48. Oakberg, E. F. and Clark, E., Effect of dose and dose rate on radiation damage to mouse spermatogonia and oocytes as measured by cell survival, *J. Cell. Comp. Physiol.*, 58, Suppl., 1, 173, 1961.

49. Proceedings, Strahlenschutz den bevolkerung bei einer nuklearkatostrophe, Interlaken, June 1968.

50. The Control of Radiation Hazards in the United States, National Advisory Committee on Radiation (1959).

51. Radioactive Contamination of the Environment: Public Health Action, National Advisory Committee on Radiation (1962).

52. Protecting and Improving Health through the Radiological Sciences, National Advisory Committee on Radiation (1966).

53. Radiation Quantities and Units (ICRU Report No. 10A) NBS, 14-84, U.S. Government Printing Office, Washington, D.C., 1962.

54. Medical Research Council., Hazards to man of nuclear and allied radiation, Appendix K. Cmd. 1225, HMSO, 1960.

Reports of the National Council on Radiation Protection and Measurements

55. X-ray Protection, NCRP Report No. 1, (NBS-HB15), 1931.

56. Radium Protection, NCRP Report No. 2, (NBS-HB-18), 1934.

57. X-ray Protection, NCRP Report No. 3, (NBS-HB-20), 1936.

58. Radium Protection, NCRP Report No. 4, (NBS-HB-23), 1938.

59. Safe Handling of Radioactive Luminous Compounds, NCRP Report No. 5, (NBS-HB-27), 1941.

60. Medical X-ray Protection up to Two Million Volts, NCRP Report No. 6, (NBS-HB 41), 1949.

61. Safe Handling of Radioactive Isotopes, NCRP Report No. 7, (NBS-HB-42), 1949.

62. Control and Removal of Radioactive Contamination in Laboratories, NCRP Report No. 8, (NBS-HB-48), 1951.

63. Recommendations for Waste Disposal of Phosphorous-32 and Iodine-131 for Medical Users, NCRP Report No. 9, (NBS-HB-49), 1951.

64. Radiological Monitoring Methods and Instruments, NCRP Report No. 10, (NBS-HB-51), 1952.

65. Maximum Permissible Amounts of Radioisotopes in the Human Body and Maximum Permissible Concentrations in Air and Water, NCRP Report No. 11, (NBS-HB-52), 1953.

66. Recommendations for the Disposal of Carbon-14 Wastes, NCRP Report No. 12, (NBS-HB-53), 1953.

67. Protection Against Radiations from Radium, Cobalt-60 and Cesium-137, NCRP Report No. 13, (NBS-HB-54), 1954.

*NCRP Report Nos. 1–28 and 30–31 were published as National Bureau of Standards Handbooks by the U.S. Government Printing Office. Reports No. 32 on are published by the National Council on Radiation Protection, Washington, D. C.

68. Protection Against Betatron-Synchrotron Radiations Up to 100 Million Volts, NCRP Report No. 14, (NBS-HB-55), 1954.

69. Safe Handling of Cadavers Containing Radioactive Isotopes, NCRP Report No. 15, (NBS-HB-55), 1953.

70. Radioactive-Waste Disposal in the Ocean, NCRP Report No. 16, (NBS-HB-58), 1954.

71. Permissible Dose from External Sources of Ionizing Radiation, NCRP Report No. 17, (NBS-HB-59), 1954.

72. X-ray Protection, NCRP Report No. 18, (NBS-HB-60), 1955.

73. Regulation of Radiation Exposure by Legislative Means, NCRP Report No. 19, (NBS-HB-61), 1955.

74. Maximum Permissible Radiation Exposure to Man—A Preliminary Statement of the National Committee on Radiation Protection and Measurements (January 8, 1957), *Amer. J. Roentgen.*, 77, 910, 1957, and *Radiology*, 68, 260, 1957.

75. Protection Against Neutron Radiation Up to 30 Million Electron Volts, NCRP Report No. 20, (NBS-HB-63), 1957.

76. Safe Handling of Bodies Containing Radioactive Isotopes, NCRP Report No. 21, (NBS-HB-65), 1958.

77. Maximum Permissible Radiation Exposures to Man (April 15, 1958), Addendum to NCRP Report No. 17, *Radiology*, 71, 263, 1958.

78. Maximum Permissible Body Burdens and Maximum Permissible Concentrations of Radionuclides in Air and in Water for Occupational Exposure, NCRP Report No. 22, (NBS-HB-69), 1959.

79. Somatic Radiation Dose for the General Population—Report of the Ad Hoc Committee of the National Committee on Radiation Protection and Measurements (May 6, 1959), *Science*, 131, 482, 1960. Copyright 1960 by the American Association for the Advancement of Science.

80. Statements on Maximum Permissible Dose from Television Receivers and Maximum Permissible Dose to the Skin of the Whole Body, *Radiology*, 75, 122, 1960.

81. Measurements of Neutron Flux and Spectra for Physical and Biological Applications, NCRP Report No. 23, (NBS-HB-72), 1960.

82. Protection Against Radiations from Sealed Gamma Sources, NCRP Report No. 24, (NBS-HB-73), 1960.

83. Measurement of Absorbed Dose of Neutrons and of Mixtures of Neutrons and Gamma Rays, NCRP Report No. 25, (NBS-HB-75), 1961.

84. Medical X-ray Protection Up to Three Million Volts, NCRP Report No. 26, (NBS-HB-76), 1961.

85. Stopping Powers for Use with Cavity Chambers, NCRP Report No. 27, (NBS-HB-79), 1961.

86. A Manual of Radioactivity Procedures, NCRP Report No. 28, (NBS-HB-80), 1961.

87. Exposure to Radiation in an Emergency, NCRP Report No. 29, (University of Chicago, Chicago, Ill., 1962)

88. Safe Handling of Radioactive Materials, NCRP Report No. 30, (NBS-HB-92), 1964.

89. Shielding for High-Energy Electron Accelerator Installations, NCRP Report No. 31, (NBS-HB-97), 1964.

90. Radiation Protection in Educational Institutions, NCRP Report No. 32, 1966.

91. Medical X-Ray and Gamma-Ray Protection For Energies Up To 10 MeV—Equipment Design and Use, Report No. 33, 1968.

92. X-Ray Protection Standards for Home Television Receivers, Interim Statement of the National Council on Radiation Protection and Measurements, (National Council on Radiation Protection and Measurements, Washington, 1968)

93. Medical X-Ray and Gamma-Ray Protection for Energies Up to 10 MeV—Structural Shielding Design and Evaluation, NCRP Report No. 34, 1970.

94. Dental X-Ray Protection, NCRP Report No. 35, 1970.

95. Radiation Protection in Veterinary Medicine, NCRP Report No. 36, 1970.

96. Precautions in the Management of Patients Who Have Received Therapeutic Amounts of Radionuclides, NCRP Report No. 37, 1970.

97. Protection Against Neutron Radiation, NCRP Report No. 38, 1971.

98. Basic Radiation Protection Criteria, NCRP Report No. 39, 1971.

Publications of the International Commission on Radiological Protection

99. X-ray and Radium Protection, Recommendations of the Second International Congress of Radiology, *Brit. J. Radiol.*, 1 (New Series), 359, 1928.

100. International Recommendations for X-ray and Radium Protection, 1931. Revised by the International X-ray and Radium Protection Committee and Adopted by the Third International Congress of Radiology, *Radiology*, 21, 212, 1933.

101. International Recommendations for X-ray and Radium Protection, 1934, *Radiology*, 23, 682, 1934.

102. International Recommendations for X-ray and Radium Protection, 1937, *Radiology*, 30, 511, 1938.

103. International Recommendations on Radiological Protection, 1950, *Radiology*, 56, 431, 1951.

104. Recommendations of the International Commission on Radiological Protection (Revised December 1, 1954), *British J. Radiol.*, Supplement No. 6, British Institute of Radiology, London, 1955.

105. Report on 1956 Amendments to the Recommendations of the International Commission on Radiological Protection (ICRP), *Radiology*, 70, 261, 1958.

106. Exposure of Man to Ionizing Radiation Arising from Medical Procedures, An Inquiry into Methods of Evaluation, A Report of the International Commission on Radiological Protection and the International Commission on Radiological Units and Measurements, *Phys. Med. Biol.* 2, 107, 1957.

107. Recommendations of the International Commission on Radiological Protection (Adopted September 9, 1958), ICRP Pub. 1, Pergamon Press, London, 1959.

108. Recommendations of the International Commission on Radiological Protection (ICRP Pub. 2), Report of Committee II on Permissible Dose for Internal Radiation (1959), *Health Phys.*, 3, 1, 1960.

109. Recommendations of the International Commission on Radiological Protection Committee III on Protection Against X-rays up to Energies of 3 MeV and Beta- and Gamma-Rays from Sealed Sources (1960), Pergamon Press, London, 1960.

110. Exposure of Man to Ionizing Radiation Arising from Medical Procedures with Special Reference to Radiation Induced Diseases—An Inquiry Into Methods of Evaluation, A Report of the International Commission on Radiological Protection and the International Commission on Radiological Units and Measurements, *Phys. Med. Biol.* 6, 199, 1961.

111. Report of the RBE Committee to the International Commissions on Radiological Protection and on Radiological Units and Measurements, *Health Phys,*, 9, 357, 1963.

112. Recommendations of the International Commission on Radiological Protection (ICRP Pub. 4), Report of Committee IV (1953-1959) on Protection Against Electromagnetic Radiation Above 3 MeV and Electrons, Neutrons and Protons (Adopted 1962, with revisions adopted 1963), Pergamon Press, Oxford, 1964.

113. Recommendations of the International Commission on Radiological Protection (ICRP Pub. 5), Report of Committee V on the Handling and Disposal of Radioactive Materials in Hospitals and Medical Research Establishments, Pergamon Press, Oxford, 1965.

114. Recommendations of the International Commission on Radiological Protection (ICRP Pub. 6), (As Amended 1959 and Revised 1962), Pergamon Press, Oxford, 1964.

115. Principles of Environmental Monitoring Related to the Handling of Radioactive Materials (ICRP Pub. 7), A Report Prepared by a Task Group of Committee 4 of the International Commission on Radiological Protection, Pergamon Press, Oxford, 1966.

116. The Evaluation of Risks from Radiation (ICRP Pub. 8), A Report Prepared by a Task Group of Committee 1 of the International Commission on Radiological Protection, *Health Phys.*, 12, 239, 1966.

117. Recommendations of the International Commission on Radiological Protection (ICRP Pub. 9), (Adopted September 17, 1965), Pergamon Press, Oxford, 1966.

118. Recommendations of the International Commission on Radiological Protection (ICRP Pub. 10), Report of Committee 4 on Evaluation of Radiation Doses to Body Tissues from Internal Contamination Due to Occupational Exposure, Pergamon Press, Oxford, 1968.

119. A Review of the Radiosensitivity of the Tissues in Bone (ICRP Pub. 11), A Report Prepared for Committees 1 and 2 of the International Commission on Radiological Protection, Pergamon Press, Oxford, 1968.

120. General Principles of Monitoring for Radiation Protection of Workers (ICRP Pub. 12), A Report by Committee 4 of the International Commission on Radiological Protection, Pergamon Press, Oxford, 1969.

121. Radiation Protection in Schools for Pupils up to Age of 18 years (ICRP Pub. 13), A Report by Committee 3 of the International Commission on Radiological Protection, Pergamon Press, Oxford, 1970.

122. Radiosensitivity and Spatial Distribution of Dose (ICRP Pub. 14), Reports Prepared by Two Task Groups of Committee 1 of the International Commission on Radiological Protection, Pergamon Press, Oxford, 1969.

Publications of the United Nations Scientific Committee on the Effects of Atomic Radiation

123. Report of the United Nations Scientific Committee on the Effects of Atomic Radiation, General Assembly Official Records: Thirteenth Session, Supplement No. 17 (A/3838) (United Nations, New York, 1958).

124. Report of the United Nations Scientific Committee on the Effects of Atomic Radiation, General Assembly Official Records: Seventeenth Session, Supplement No. 16 (A/5216) (United Nations, New York, 1962).

125. Report of the United Nations Scientific Committee on the Effects of Atomic Radiation, General Assembly Official Records: Nineteenth Session, Supplement No. 14 (A/5814) (United Nations, New York, 1964).

126. Report of the United Nations Scientific Committee on the Effects of Atomic Radiation, General Assembly Official Records: Twenty-first Session, Supplement No. 14 (A/6314) (United Nations, New York, 1966).

Publications of the Federal Radiation Council*

127. Background Material for the Development of Radiation Protection Standards (FRC Report No. 1, 1960).

128. Federal Radiation Council, Radiation Protection Guidance for Federal Agencies, Memorandum for the President, Federal Register, May 18, 1960.

129. Background Material for the Development of Radiation Protection Standards (FRC Report No. 2, 1961).

130. Federal Radiation Council, Radiation Protection Guidance for Federal Agencies, Memorandum for the President, Federal Register, September 26, 1961.

*U. S. Government Printing Office, Washington, D. C.

131. Health Implications of Fallout From Nuclear Weapons Testing through 1961 (FRC Report No. 3, 1962).

132. Pathological Effects of Thyroid Irradiation, A Report of a Panel of Experts from the Committees on the Biological Effects of Atomic Radiation: National Academy of Sciences - National Research Council (Federal Radiation Council, Washington, 1962).

133. Estimates and Evaluation of Fallout in the United States from Nuclear Weapons Testing Conducted Through 1962 (FRC Report No. 4, 1963).

134. Background Material for the Development of Radiation Protection Standards (FRC Report No. 5, 1964).

135. Federal Radiation Council, Radiation Protection Guidance for Federal Agencies, Memorandum for the President Federal Register, August 22, 1964.

136. Revised Fallout Estimates for 1964-1965 and Verification of the 1963 Prediction (FRC Report No. 6, 1964).

137. Implications to Man of Irradiation by Internally Deposited Strontium-89, Strontium-90, and Cesium-137, A Report of an Advisory Committee from the Division of Medical Sciences: National Academy of Sciences – National Research Council (Federal Radiation Council, Washington, 1964).

138. Background Material for the Development of Radiation Protection Standards – Protective Action Guides for Strontium-89, Strontium-90 and Cesium-137, (ICRP Report No. 7, 1965).

139. Federal Radiation Council, Radiation Protection Guidances for Federal Agencies, Memorandum for the President, Federal Register, May 22, 1965.

140. Pathological Effects of Thyroid Irradiation – Revised Report, A Report of an Advisory Committee from the Division of Medical Sciences: National Research Council (Federal Radiation Council, Washington, 1966).

141. Guidance for the Control of Radiation Hazards in Uranium Mining, (ICRP Report No. 8, 1967).

142. Radiation Exposure of Uranium Miners – Report of an Advisory Committee from the Division of Medical Sciences: National Academy of Sciences – National Research Council – National Academy of Engineering (Federal Radiation Council, Washington, 1968).

Publications of the Joint Committee on Atomic Energy*

143. The Nature of Radioactive Fallout and Its Effects on Man, Part 1 – Hearings Before the Special Subcommittee on Radiation, May 27, 28, 29, and June 3, 1957.

144. The Nature of Radioactive Fallout and Its Effects on Man, Part 2 – Hearings Before the Special Subcommittee on Radiation, June 4, 5, 6, and 7, 1957.

145. The Nature of Radioactive Fallout and Its Effects on Man, Part 3: Index – Hearings Before the Special Subcommittee on Radiation, May 27, 28, 29, June 3, 4, 5, 6, and 7, 1957.

146. The Nature of Radioactive Fallout and Its Effects on Man – Summary – Analysis of Hearings May 27-29 and June 3-7, 1957, Before the Special Subcommittee on Radiation.

147. Industrial Radioactive Waste Disposal, Volumes 1-4 – Hearings Before the Special Subcommittee on Radiation, January 28, 29, 30, February 2 and 3, 1959.

148. Industrial Radioactive Waste Disposal, Volume 5 – Hearings Before the Special Subcommittee on Radiation, July 29, 1959.

149. Industrial Radioactive Waste Disposal – Summary – Analysis of Hearings January 28, 29, 30, February 2, 3, and July 29, 1959, Before the Special Subcommittee on Radiation.

150. Selected Materials on Employee Radiation Hazards and Workmen's Compensation.

*U. S. Government Printing Office, Washington, D. C.

151. Employee Radiation Hazards and Workmen's Compensation – Hearings Before the Subcommittee on Research and Development March 10, 11, 12, 17, 18, and 19, 1959.

152. Employee Radiation Hazards and Workmen's Compensation – Summary – Analysis of Hearings March 10, 11, 12, 17, 18, and 19, 1959, Before the Subcommittee on Research and Development.

153. Fallout from Nuclear Weapons Tests, Volumes 1-3 – Hearings Before the Special Subcommittee on Radiation, May 5, 6, 7, and 8, 1959.

154. Fallout from Nuclear Weapons Tests – Summary – Analysis of Hearings May 5-8, 1959, Before the Special Subcommittee on Radiation.

155. Selected Materials on Federal-State Cooperation in the Atomic Energy Field, 1959.

156. Federal-State Relationships in the Atomic Energy Field, 1959.

157. Federal-State Relationships in the Atomic Energy Field – Hearings Before the Joint Committee on Atomic Energy, May 19, 20, 21, 22, and August 26, 1959.

158. Selected Materials on Radiation Protection Criteria and Standards: Their Basis and Use, 1960.

159. Radiation Protection Criteria and Standards: Their Basis and Use – Hearings Before the Special Subcommittee on Radiation, May 24, 25, 26, 31, June 1, 2, and 3, 1960.

160. Radiation Protection Criteria and Standards: Their Basis and Use – Summary-Analysis of Hearings May 24, 25, 26, 31, June 1, 2, and 3, 1960, Before Special Subcommittee on Radiation.

161. Radiation Safety and Regulation – Hearings Before the Joint Committee on Atomic Energy, June 12, 13, 14, and 15, 1961.

162. Radiation Standards, Including Fallout, Parts 1 and 2 – Hearings Before the Subcommittee on Research, Development and Radiation, June 4, 5, 6, and 7, 1962.

163. Radiation Standards, Including Fallout – Summary – Analysis of Hearings Before the Subcommittee on Research, Development and Radiation, June 4, 5, 6, and 7, 1962.

164. Fallout, Radiation Standards, and Countermeasures Part 1 – Hearings Before the Subcommittee on Research, Development, and Radiation, June 3, 4, and 6, 1963.

165. Fallout, Radiation Standards and Countermeasures Part 2 – Hearings Before the Subcommittee on Research, Development, and Radiation, August 20, 21, 22, and 27, 1963.

166. Federal Radiation Council Protective Action Guides – Hearings Before the Subcommittee on Research, Development, and Radiation, June 29, and 30, 1965.

167. Proposed Legislation Relating to Uniform Record Keeping and Workmen's Compensation Coverage for Radiation Workers – Hearings Before the Joint Committee on Atomic Energy, August 30, 31, September 20, 21, and 22, 1966.

168. Radiation Exposure of Uranium Miners, Part 1 – Hearings Before the Subcommittee on Research, Development and Radiation, May 9, 10, 23, June 6, 7, 8, 9, July 26, 27, and August 8, 10, 1967.

169. Radiation Exposure of Uranium Miners, Part 2 – Additional Backup and Reference Material.

170. Radiation Exposure of Uranium Miners – Summary Analysis of Hearings May 9, 10, 23, June 6, 7, 8, 9, July 26, 27, and August 8, 10, 1967, Before the Subcommittee on Research, Development and Radiation.

Publication of the National Academy of Sciences - National Research Council*

171. The Biological Effects of Atomic Radiation – Summary Reports, 1956.

*National Academy of Sciences-National Research Council, Washington, D. C.

172. The Biological Effects of Atomic Radiation – A Report to the Public, 1956.

173. Pathologic Effects of Atomic Radiation, Pub. 452, 1956.

174. A Commentary on the Report of the United Nations Scientific Committee on the Effects of Atomic Radiation – Report II of the Committee on Pathologic Effects of Atomic Radiation, Pub. 647, 1959.

175. Radioactive Waste Disposal into Atlantic and Gulf Coastal Waters, Pub. 655, 1959.

176. Radioactive Waste Disposal from Nuclear Powered Ships, Pub. 658, 1959.

177. The Biological Effects of Atomic Radiation – Summary Reports, 1960.

178. Effects of Inhaled Radioactive Particles, Pub. 848, 1961.

179. Long-Term Effects of Ionizing Radiations from External Sources, Pub. 849, 1961.

180. Effects of Ionizing Radiation on the Human Hemapoietic System, Pub. 875, 1961.

181. Internal Emitters, Pub. 883, 1961.

182. Disposal of Low-Level Radioactive Waste Into Pacific Coastal Water, Pub. 985, 1962.

183. The Behavior of Radioactive Fallout in Soils and Plants, Pub. 1092, 1963.

184. Radiobiological Factors in Manned Space Flight, Pub. 1487, 1967.

Miscellaneous Publications

185. Safety Code for the Industrial Use of X-rays (American War Standard Z54.1), Sectional Committee Z54 of the American Standards Association (American Standards Association, New York, 1946).

186. Radiation Protection in Uranium Mines and Mills (Concentrators) – American Standard N7.1-1960, Sectional Committee N7 of the American Standards Association (American Standards Association, New York, 1961).

187. Cowie, D. B. and Scheele., L. A., A Survey of Radiation Protection in Hospitals, *J. Nat. Cancer Inst.*, 1, 1941.

188. Radiation Control for Health and Safety Act of 1967 – Hearings Before the Committee on Commerce, United States Senate, August 28, 29, and 30, 1967 (U. S. Government Printing Office, Washington, D. C., 1968).

189. Radiation Control for Health and Safety Act of 1967, Part 2 – Hearings Before the Committee on Commerce, United States Senate. May 6, 8, 9, 13, and 15, 1968 (U. S. Government Printing Office, Washington, D. C., 1968)

614·876

LIBRARY
No. B5630
27 OCT 1972
R.P.E. WESTCOTT